Two Manuals Combined into One Book



# Military First Aid

# And

# Soldier's Handbook for Individual Operations Survival in Cold Weather Areas

Department of Defense

Pentagon
Publishing

*FIELD MANUAL
NO. 4-25.11
NAVY TACTICAL
REFERENCE
PUBLICATION
NO. 4-02.1
AIR FORCE MANUAL
NO. 44-163(I)

HEADQUARTERS
DEPARTMENT OF THE ARMY,
THE NAVY, AND THE AIR FORCE
Washington, DC, 23 December 2002

# FIRST AID

## TABLE OF CONTENTS

Page



*This publication supersedes FM 21-11, 27 October 1988

# PREFACE

This manual meets the first aid training needs of individual service members. Because medical personnel will not always be readily available, the nonmedical service members must rely heavily on their own skills and knowledge of life-sustaining methods to survive on the integrated battlefield. This publication outlines both self-aid and aid to other service members (buddy aid). More importantly, it emphasizes prompt and effective action in sustaining life and preventing or minimizing further suffering and disability. First aid is the emergency care given to the sick, injured, or wounded before being treated by medical personnel. The term *first aid* can be defined as "urgent and immediate lifesaving and other measures, which can be performed for casualties by nonmedical personnel when medical personnel are not immediately available." Nonmedical service members have received basic first aid training and should remain skilled in the correct procedures for giving first aid. This manual is directed to *all* service members. The procedures discussed apply to all types of casualties and the measures described are for use by both male and female service members.

This publication is in consonance with the following North Atlantic Treaty Organization (NATO) International Standardization Agreements (STANAGs) and American, British. Canadian, and Australian Quadripartite Standardization Agreements (QSTAGs).

| TITLE | STANAG | QSTAG |
|---|---|---|
| Medical Training in First Aid, Basic Hygiene and Emergency Care | 2122 | 535 |
| First Aid Kits and Emergency Medical Care Kits | 2126 | |
| Medical First Aid and Hygiene Training in NBC Operations | 2358 | |
| First Aid Material for Chemical Injuries | 2871 | |

These agreements are available on request, using Department of Defense (DD) Form 1425 from the Standardization Documents Order Desk, 700 Robins Avenue, Building 4, Section D, Philadelphia, Pennsylvania 19111-5094.

Unless this publication states otherwise, masculine nouns and pronouns do not refer exclusively to men.

Use of trade or brand names in this publication is for illustrative purposes only and does not imply endorsement by the Department of Defense (DOD).

The proponent for this publication is the US Army Medical Department Center and School. Submit comments and recommendations for the improvement of this publication directly to the **Commander, US Army Medical Department Center and School, ATTN: MCCS-FCD-L, 1400 East Grayson Street, Fort Sam Houston, Texas 78234-5052**.

CHAPTER 1

# FUNDAMENTAL CRITERIA FOR FIRST AID

**"The fate of the wounded rests in the hands of the ones who apply the first dressing."**

*Nicholas Senn (1898) (49th President of the American Medical Association)*

## 1-1. General

When a nonmedical service member comes upon an unconscious or injured service member, he must accurately evaluate the casualty to determine the first aid measures needed to prevent further injury or death. He should seek medical assistance as soon as possible, but he should not interrupt the performance of first aid measures. To interrupt the first aid measures may cause more harm than good to the casualty. Remember that in a chemical environment, the service member should not evaluate the casualty until the casualty has been masked. After performing first aid, the service member must proceed with the evaluation and continue to monitor the casualty for development of conditions which may require the performance of necessary basic lifesaving measures, such as clearing the airway, rescue breathing, preventing shock, and controlling bleeding. He should continue to monitor the casualty until relieved by medical personnel.

Service members may have to depend upon their first aid knowledge and skills to save themselves (self-aid) or other service members (buddy aid/ combat lifesaver). They may be able to save a life, prevent permanent disability, or reduce long periods of hospitalization by knowing **WHAT** to do, **WHAT NOT** to do, and **WHEN** to seek medical assistance.

**NOTE**

The prevalence of various body armor systems currently fielded to US service members, and those in development for future fielding, may present a temporary obstacle to effective evaluation of an injured service member. You may have to *carefully remove* the body armor from the injured service member to complete the evaluation or administer first aid. Begin by removing the outer-most hard or soft body armor components (open, unfasten or cut the closures, fasteners, or straps), then remove any successive layers in the same manner. Be sure to follow other notes, cautions and warnings regarding procedures in contaminated situations and when a broken back or neck is suspected. Continue to evaluate.

## 1-2. Terminology

To enhance the understanding of the material contained in this publication, the following terms are used—

- *Combat lifesaver.* This is a US Army program governed by Army Regulation (AR) 350-41. The combat lifesaver is a member of a nonmedical unit selected by the unit commander for additional training beyond basic first aid procedures (referred to as *enhanced first aid*). A minimum of one individual per squad, crew, team, or equivalent-sized unit should be trained. The primary duty of this individual does not change. The additional duty of combat lifesaver is to provide enhanced first aid for injuries based on his training before the trauma specialist (military occupational specialty [MOS] 91W) arrives. The combat lifesaver's training is normally provided by medical personnel assigned, attached, or in direct support (DS) of the unit. The senior medical person designated by the commander manages the training program.

- *Trauma Specialist (US Army) or Hospital Corpsman (HM).* A medical specialist trained in emergency medical treatment (EMT) procedures and assigned or attached in support of a combat or combat support unit or marine forces.

- *Casualty evacuation.* Casualty evacuation (CASEVAC) is a term used by nonmedical units to refer to the movement of casualties aboard nonmedical vehicles or aircraft. See also the term *transported* below. Refer to FM 8-10-6 for additional information.

**CAUTION**

Casualties transported in this manner do not receive en route medical care.

- *Enhanced first aid (US Army).* Enhanced first aid is administered by the combat lifesaver. It includes measures, which require an additional level of training above self-aid and buddy aid, such as the initiation of intravenous (IV) fluids.

- *Medical evacuation.* Medical evacuation is the timely, efficient movement of the wounded, injured, or ill service members from the battlefield and other locations to medical treatment facilities (MTFs). Medical personnel provide en route medical care during the evacuation. Once the casualty has entered the medical stream (trauma specialist, hospital corpsman, evacuation crew, or MTF), the role of first aid in the care of the casualty ceases and the casualty becomes the responsibility of the health service support (HSS) chain. Once he has entered the HSS chain he is referred to as a *patient*.

- *First aid measures*. Urgent and immediate lifesaving and other measures, which can be performed for casualties (or performed by the casualty himself) by nonmedical personnel when medical personnel are not immediately available.

- *Medical treatment*. Medical treatment is the care and management of wounded, injured, or ill service members by medically trained (MOS-trained) HM, and area of concentration (AOC) personnel. It may include EMT, advanced trauma management (ATM), and resuscitative and surgical intervention.

- *Medical treatment facility*. Any facility established for the purpose of providing medical treatment. This includes battalion aid stations, Level II facilities, dispensaries, clinics, and hospitals.

- *Self-aid/buddy aid*. Each individual service member is trained to be proficient in a variety of specific first aid procedures. This training enables the service member or a buddy to apply immediate first aid measures to alleviate a life-threatening situation.

- *Transported*. A casualty is moved to an MTF in a nonmedical vehicle without en route care provided by a medically-trained service member (such as a Trauma Specialist or HM). First aid measures should be continually performed while the casualty is being transported. If the casualty is acquired by a dedicated medical vehicle with a medically-trained crew, the role of first aid ceases and the casualty becomes the responsibility of the HSS chain, and is then referred to as a *patient*. This method of transporting a casualty is also referred to as *CASEVAC*.

## 1-3. Understanding Vital Body Functions for First Aid

In order for the service member to learn to perform first aid procedures, he must have a basic understanding of what the vital body functions are and what the result will be if they are damaged or not functioning.

*a. Breathing Process*. All humans must have oxygen to live. Through the breathing process, the lungs draw oxygen from the air and put it into the blood. The heart pumps the blood through the body to be used by the cells that require a constant supply of oxygen. Some cells are more dependent on a constant supply of oxygen than others. For example, cells of the brain may die within 4 to 6 minutes without oxygen. Once these cells die, they are lost forever since they do not regenerate. This could result in permanent brain damage, paralysis, or death.

*b. Respiration.* Respiration occurs when a person inhales (oxygen is taken into the body) and then exhales (carbon dioxide [$CO_2$] is expelled from the body). Respiration involves the—

- *Airway.* The airway consists of the nose, mouth, throat, voice box, and windpipe. It is the canal through which air passes to and from the lungs.

- *Lungs.* The lungs are two elastic organs made up of thousands of tiny air spaces and covered by an airtight membrane. The *bronchial tree* is a part of the lungs.

- *Rib cage.* The rib cage is formed by the muscle-connected ribs, which join the spine in back, and the breastbone in front. The top part of the rib cage is closed by the structure of the neck, and the bottom part is separated from the abdominal cavity by a large dome-shaped muscle called the *diaphragm* (Figure 1-1). The diaphragm and rib muscles, which are under the control of the respiratory center in the brain, automatically *contract* and *relax*. *Contraction* increases and *relaxation* decreases the size of the rib cage. When the rib cage increases and then decreases, the air pressure in the lungs is first less and then more than the atmospheric pressure, thus causing the air to rush into and out of the lungs to equalize the pressure. This cycle of inhaling and exhaling is repeated about 12 to 18 times per minute.

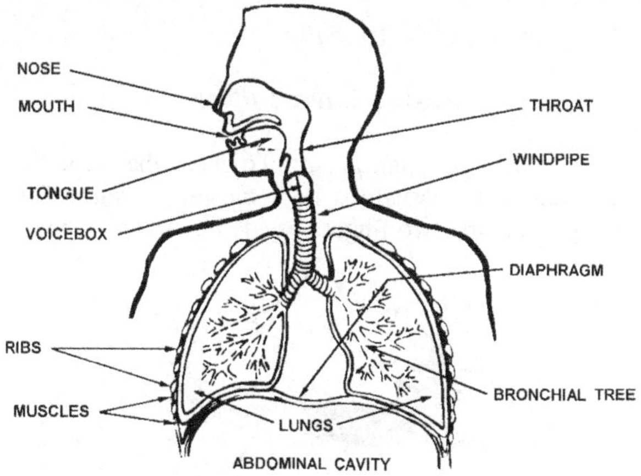


*Figure 1-1. Airway, lungs, and rib cage.*

*c. Blood Circulation.* The heart and the blood vessels (arteries, veins, and capillaries) circulate blood through the body tissues. The heart is divided into two separate halves, each acting as a pump. The left side pumps oxygenated blood (bright red) through the arteries into the capillaries; nutrients and oxygen pass from the blood through the walls of the capillaries into the cells. At the same time waste products and $CO_2$ enter the capillaries. From the capillaries the oxygen poor blood is carried through the veins to the right side of the heart and then into the lungs where it expels the $CO_2$ and picks up oxygen. Blood in the veins is dark red because of its low oxygen content. Blood does not flow through the veins in spurts as it does through the arteries. The entire system of the heart, blood vessels, and lymphatics is called the *circulatory system.*

(1) *Heartbeat.* The heart functions as a pump to circulate the blood continuously through the blood vessels to all parts of the body. It contracts, forcing the blood from its chambers; then it relaxes, permitting its chambers to refill with blood. The rhythmical cycle of contraction and relaxation is called the *heartbeat.* The normal heartbeat is from 60 to 80 beats per minute.

(2) *Pulse.* The heartbeat causes a rhythmical expansion and contraction of the arteries as it forces blood through them. This cycle of expansion and contraction can be felt (monitored) at various points in the body and is called the *pulse.* The common points for checking the pulse are at the—

- Side of the neck (*carotid*).
- Groin (*femoral*).
- Wrist (*radial*).
- Ankle (*posterior tibial*).

(*a*) *Carotid pulse.* To check the carotid pulse, feel for a pulse on the side of the casualty's neck closest to you. This is done by placing the tips of your first two fingers beside his Adam's apple (Figure 1-2).

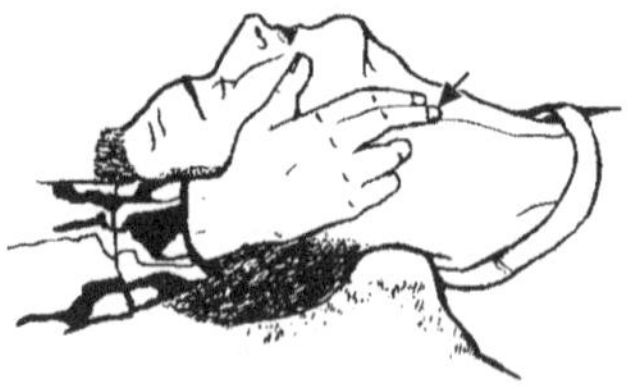

*Figure 1-2. Carotid pulse.*

(*b*) *Femoral pulse*. To check the femoral pulse, press the tips of your first two fingers into the middle of the groin (Figure 1-3).

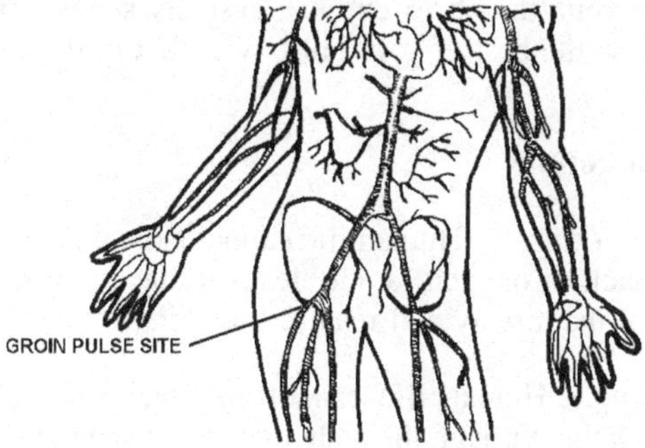


*Figure 1-3. Femoral pulse.*

(*c*) *Radial pulse*. To check the radial pulse, place your first two fingers on the thumb side of the casualty's wrist (Figure 1-4).

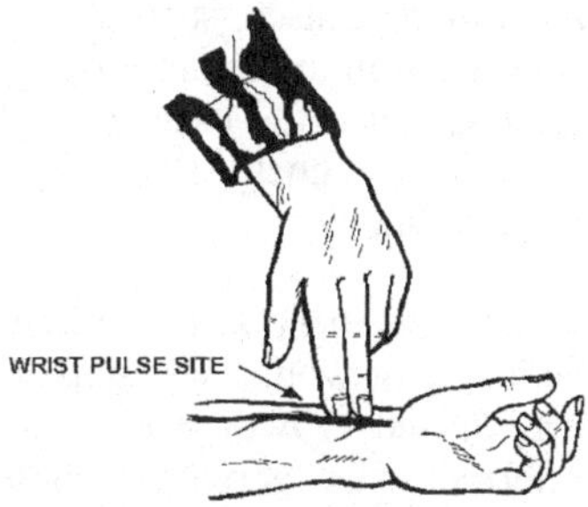


*Figure 1-4. Radial pulse.*

(*d*) *Posterior tibial pulse*. To check the posterior tibial pulse, place your first two fingers on the inside of the ankle (Figure 1-5).

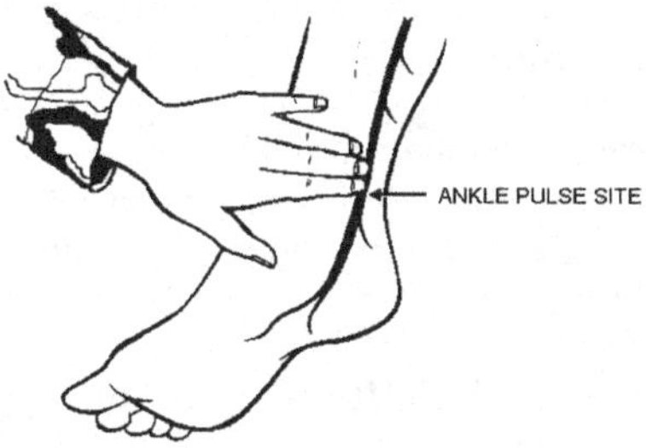


*Figure 1-5. Posterior tibial pulse.*

NOTE

**DO NOT** use your thumb to check a casualty's pulse because you may confuse the beat of your pulse with that of the casualty.

## 1-4. Adverse Conditions

*a. Lack of Oxygen.* Human life cannot exist without a continuous intake of oxygen. Lack of oxygen rapidly leads to death. First aid involves knowing how to open the airway and restore breathing.

*b. Bleeding.* Human life cannot continue without an adequate volume of blood circulating through the body to carry oxygen to the tissues. An important first aid measure is to stop the bleeding to prevent the loss of blood.

*c. Shock.* Shock means there is an inadequate blood flow to the vital tissues and organs. Shock that remains uncorrected may result in death even though the injury or condition causing the shock would not otherwise be fatal. Shock can result from many causes, such as loss of blood, loss of fluid from deep burns, pain, and reaction to the sight of a wound or blood. First aid includes preventing shock, since the casualty's chances of survival are much greater if he does not develop shock. Refer to paragraphs 2-21 through 2-24 for a further discussion of shock.

*d. Infection.* Recovery from a severe injury or a wound depends largely upon how well the injury or wound was initially protected. Infections result from the multiplication and growth (spread) of harmful microscopic organisms (sometimes referred to as germs). These harmful microscopic organisms are in the air, water, and soil, and on the skin and clothing. Some of these organisms will immediately invade (contaminate) a break in the skin or an open wound. The objective is to keep wounds clean and free of these organisms. A good working knowledge of basic first aid measures also includes knowing how to dress a wound to avoid infection or additional contamination.

## 1-5. Basics of First Aid

Most injured or ill service members are able to return to their units to fight or support primarily because they are given appropriate and timely first aid followed by the best medical care possible. Therefore, all service members must remember the basics.

- Check for **BREATHING**: Lack of oxygen intake (through a compromised airway or inadequate breathing) can lead to brain damage or death in very few minutes.

- Check for **BLEEDING**: Life cannot continue without an adequate volume of blood to carry oxygen to tissues.

- Check for **SHOCK**: Unless shock is prevented, first aid performed, and medical treatment provided, death may result even though the injury would not otherwise be fatal.

## 1-6. Evaluating a Casualty

*a.* The time may come when you must instantly apply your knowledge of first aid measures. This could occur during combat operations, in training situations, or while in a nonduty status. Any service member observing an unconscious and/or ill, injured, or wounded person must carefully and skillfully evaluate him to determine the first aid measures required to prevent further injury or death. He should seek help from medical personnel as soon as possible, but must not interrupt his evaluation of the casualty or fail to administer first aid measures. A second service member may be sent to find medical help. One of the cardinal principles for assisting a casualty is that you (the initial rescuer) must continue the evaluation and first aid measures, as the tactical situation permits, until another individual relieves you. If, during any part of the evaluation, the casualty exhibits the conditions (such as shock) for which the service member is checking, the service member must stop the evaluation and immediately administer first aid. In a chemical environment, the service member should not evaluate the casualty until both the individual and the casualty have been masked. If it is suspected that a nerve agent was used, administer the casualty's own nerve agent antidote autoinjector. After providing first aid, the service member must proceed with the evaluation and continue to monitor the casualty for further complications until relieved by medical personnel.

**WARNING**

**Do not use your own nerve agent antidote autoinjector on the casualty.**

**NOTE**

Remember, when evaluating and/or administering first aid to a casualty, you should seek medical aid as soon as possible. **DO NOT** stop first aid measures, but if the situation allows, send another service member to find medical aid.

*b.* To evaluate a casualty, perform the following steps:

(1) *Check the casualty for responsiveness.* This is done by gently shaking or tapping him while calmly asking, "Are you OK?" Watch for a response. If the casualty does not respond, go to step (2). If the casualty responds, continue with the evaluation.

(*a*) If the casualty is conscious, ask him where he feels different than usual or where it hurts. Ask him to identify the location of pain if he can, or to identify the area in which there is no feeling.

(*b*) If the casualty is conscious but is choking and cannot talk, stop the evaluation and begin first aid measures. Refer to paragraphs 2-10 and 2-11 for specific information on opening the airway.

**WARNING**

**If a broken back or neck is suspected, do not move the casualty unless his life is in immediate danger (such as close to a burning vehicle). Movement may cause permanent paralysis or death.**

(2) *Check for breathing.* (Refer to paragraph 2-6 for this procedure.)

(*a*) If the casualty is breathing, proceed to step (3).

(*b*) If the casualty is not breathing, stop the evaluation and begin first aid measures to attempt to ventilate the casualty. Attempt to open the airway, if an airway obstruction is apparent, clear the airway obstruction, then ventilate (see paragraphs 2-10 and 2-11).

(*c*) After successfully ventilating the casualty, proceed to step (3).

(3) *Check for pulse.* (Refer to paragraph 1-3*c*(2) for specific methods.) If a pulse is present and the casualty is breathing, proceed to step (4).

(*a*) If a pulse is present, but the casualty is still not breathing, start rescue breathing.

(*b*) If a pulse is not present, seek medical personnel for help.

(4) *Check for bleeding.* Look for spurts of blood or blood-soaked clothes. Also check for *both* entry and exit wounds. If the casualty is bleeding from an open wound, stop the evaluation and begin first aid procedures as follows for a—

(*a*) Wound of the arm or leg (refer to paragraphs 2-16 through 2-18 for information on putting on a field or pressure dressing).

(*b*) Partial or complete amputation, apply dressing (refer to paragraph 2-16 to 2-18) and then apply tourniquet if bleeding is not stopped (refer to paragraph 2-20 for information on putting on a tourniquet).

(*c*) Open head wound (refer to paragraph 3-10 for information on applying a dressing to an open head wound).

(*d*) Open chest wound (refer to paragraph 3-5 for information on applying a dressing to an open chest wound).

(*e*) Open abdominal wound (refer to paragraph 3-7 for information on applying a dressing to an open abdominal wound).

**WARNING**

**In a chemically contaminated area, do not expose the wounds. Apply field dressing and then pressure dressing over wound area as needed.**

(5) *Check for shock.* (Refer to paragraph 2-24 for first aid measures for shock.) If the signs and symptoms of shock are present, stop the evaluation, and begin first aid measures immediately. The following are the nine signs and symptoms of shock.

(*a*) Sweaty but cool skin (clammy skin).

(*b*) Paleness of skin. (In dark-skinned service members look for a grayish cast to the skin.)

(*c*) Restlessness or nervousness.

(*d*) Thirst.

(*e*) Loss of blood (bleeding).

(*f*) Confusion (does not seem aware of surroundings).

(*g*) Faster than normal breathing rate.

(*h*) Blotchy or bluish skin, especially around the mouth.

(*i*) Nausea or vomiting.

**WARNING**

**Leg fractures must be splinted before elevating the legs as a first aid measure for shock.**

(6) *Check for fractures*.

(*a*) Check for the following signs and symptoms of a back or neck injury and perform first aid procedures as necessary.

- Pain or tenderness of the back or neck area.
- Cuts or bruises on the back or neck area.
- Inability of a casualty to move or decreased sensation to extremities (paralysis or numbness).
  - Ask about ability to move (paralysis).
  - Touch the casualty's arms and legs and ask whether he can feel your hand (numbness).
- Unusual body or limb position.

(*b*) Immobilize any casualty suspected of having a back or neck injury by doing the following:

- Tell the casualty not to move.
- If a back injury is suspected, place padding (rolled or folded to conform to the shape of the arch) under the natural arch of the casualty's back. (For example, a blanket/poncho may be used as padding.)

> **WARNING**
>
> **Do not move casualty to place padding.**

- If a neck injury is suspected, immediately immobilize (manually) the head and neck. Place a roll of cloth under the casualty's neck, and put weighted boots (filled with dirt or sand) or rocks on both sides of his head.

(*c*) Check the casualty's arms and legs for open or closed fractures.

- Check for *open* fractures by looking for—
  - Bleeding.
  - Bones sticking through the skin.
  - Check for pulse.
- Check for *closed* fractures by looking for—
  - Swelling.
  - Discoloration.
  - Deformity.
  - Unusual body position.
  - Check for pulse.

(*d*) Stop the evaluation and begin first aid measures if a fracture to an arm or leg is suspected. Refer to Chapter 4 for information on splinting a suspected fracture.

(*e*) Check for signs/symptoms of fractures of other body areas (for example, shoulder or hip) and provide first aid as necessary.

(7) *Check for burns.* Look carefully for reddened, blistered, or charred skin; also check for singed clothing. If burns are found, stop the evaluation and begin first aid procedures. Refer to paragraph 3-9 for information on giving first aid for burns.

**NOTE**

Burns to the upper torso and face may cause respiratory complications. When evaluating the casualty, look for singed nose hair, soot around the nostrils, and listen for abnormal breath sounds or difficulty breathing.

(8) *Check for possible head injury.*

(*a*) Look for the following signs and symptoms:

- Unequal pupils.
- Fluid from the ear(s), nose, mouth, or injury site.
- Slurred speech.
- Confusion.
- Sleepiness.
- Loss of memory or consciousness.
- Staggering in walking.
- Headache.
- Dizziness.
- Nausea or vomiting.
- Paralysis.
- Convulsions or twitches.
- Bruising around the eyes and behind the ears.

(*b*) If a head injury is suspected, continue to watch for signs which would require performance of rescue breathing, first aid measures for shock, or control of bleeding; seek medical aid. Refer to paragraph 3-10 for information on first aid measures for head injuries.

CHAPTER 2

# BASIC MEASURES FOR FIRST AID

### 2-1. General

Several conditions that require immediate attention are an inadequate airway, lack of breathing, and excessive loss of blood (circulation). A casualty without a clear airway or who is not breathing may die from lack of oxygen. Excessive loss of blood may lead to shock, and shock can lead to death; therefore, you must act immediately to control the loss of blood. All wounds are considered to be contaminated, since infection-producing organisms (germs) are always present on the skin and clothing, and in the soil, water, and air. Any missile or instrument (such as a bullet, shrapnel, knife, or bayonet) causing a wound pushes or carries the germs into that wound. Infection results as these organisms multiply. That a wound is contaminated does not lessen the importance of protecting it from further contamination. You must dress and bandage a wound as soon as possible to prevent further contamination.

**NOTE**

It is also important that you attend to any airway, breathing, or bleeding problems **IMMEDIATELY** because these problems, if left unattended, may become life threatening.

## Section I. OPEN THE AIRWAY AND RESTORE BREATHING

### 2-2. Breathing Process

All humans must have oxygen to live. Through the breathing process, the lungs draw oxygen from the air and put it into the blood. The heart pumps the blood through the body to be used by the cells that require a constant supply of oxygen. Some cells are more dependent on a constant supply of oxygen than others. For example, cells of the brain may die within 4 to 6 minutes without oxygen. Once these cells die, they are lost forever since they do not regenerate. This could result in permanent brain damage, paralysis, or death.

### 2-3. Assessment of and Positioning the Casualty

*a.* **CHECK** for responsiveness (Figure 2-1A)—establish whether the casualty is conscious by gently shaking him and asking, “Are you OK?”

*b.* **CALL** for help (Figure 2-1B).

*c.* **POSITION** the unconscious casualty so that he is lying on his back and on a firm surface (Figure 2-1C).

**WARNING**

**If the casualty is lying on his chest (prone position), cautiously roll the casualty as a unit so that his body does not twist (which may further complicate a back, neck, or spinal injury).**

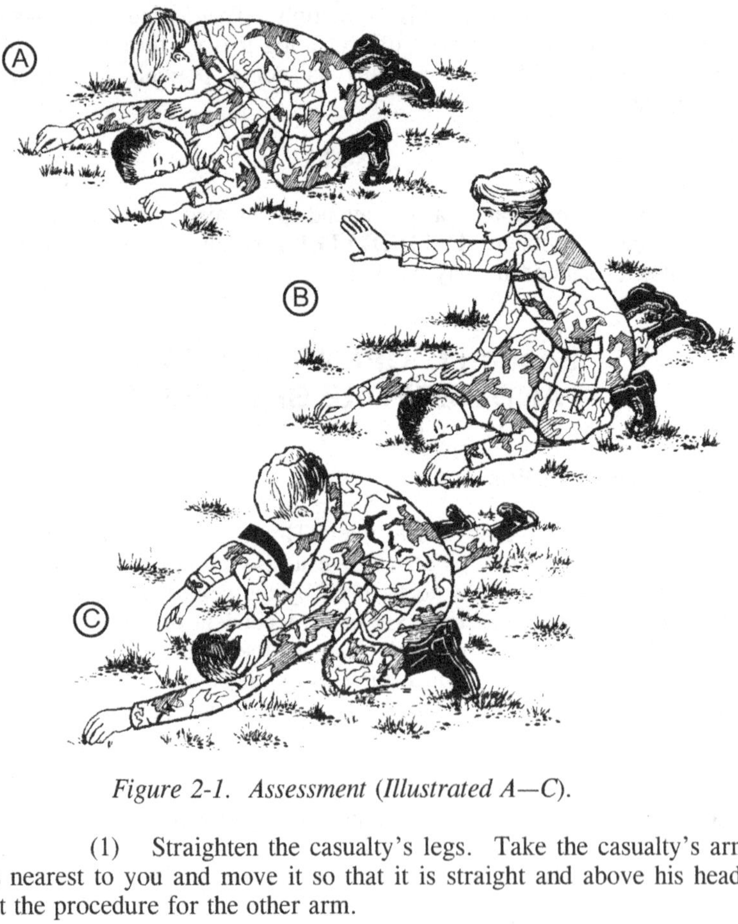


*Figure 2-1. Assessment (Illustrated A—C).*

(1) Straighten the casualty's legs. Take the casualty's arm that is nearest to you and move it so that it is straight and above his head. Repeat the procedure for the other arm.

(2) Kneel beside the casualty with your knees near his shoulders (leave space to roll his body) (Figure 2-1B). Place one hand behind his head and neck for support. With your other hand, grasp the casualty under his far arm (Figure 2-1C).

(3) Roll the casualty towards you using a steady, even pull. His head and neck should stay in line with his back.

(4) Return the casualty's arms to his side. Straighten his legs. Reposition yourself so that you are now kneeling at the level of the casualty's shoulders. However, if a neck injury is suspected and the jaw-thrust technique will be used, kneel at the casualty's head, looking towards his feet.

## 2-4. Opening the Airway of an Unconscious or Not Breathing Casualty

The tongue is the single most common cause of an airway obstruction (Figure 2-2). In most cases, simply using the head-tilt/chin-lift technique can clear the airway. This action pulls the tongue away from the air passage in the throat (Figure 2-3).

*Figure 2-2. Airway blocked by tongue.*

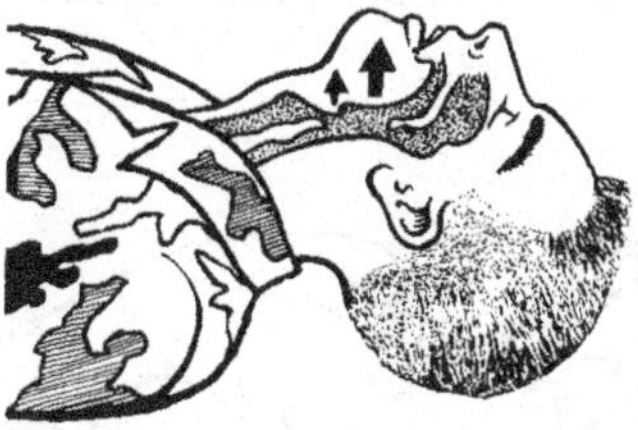

*Figure 2-3. Airway opened by extending neck.*

*a.* Call for help and then position the casualty. Move (roll) the casualty onto his back (Figure 2-1C). (Refer to paragraph 2-3*c* for information on positioning the casualty.)

**NOTE**

Perform finger sweep. If foreign material or vomitus is visible in the mouth, it should be removed, but do not spend an excessive amount of time doing so.

*b.* Open the airway using the jaw-thrust or head-tilt/chin-lift technique.

**CAUTION**

The head-tilt/chin-lift technique is an important procedure in opening the airway; however, use extreme care because excess force in performing this maneuver may cause further spinal injury. In a casualty with a suspected neck injury or severe head trauma, the safest approach to opening the airway is the jaw-thrust technique because in most cases it can be accomplished without extending the neck.

(1) *Perform the jaw-thrust technique.* The jaw-thrust may be accomplished by the rescuer grasping the angles of the casualty's lower jaw and lifting with both hands, one on each side, displacing the jaw forward and up (Figure 2-4). The rescuer's elbows should rest on the surface on which the casualty is lying. If the lips close, the lower lip can be retracted with the thumb. If mouth-to-mouth breathing is necessary, close the nostrils by placing your cheek tightly against them. The head should be carefully supported without tilting it backwards or turning it from side to side. If this is unsuccessful, the head should be tilted back very slightly. The jaw-thrust is the safest first approach to opening the airway of a casualty who has a suspected neck injury because in most cases it can be accomplished without extending the neck.

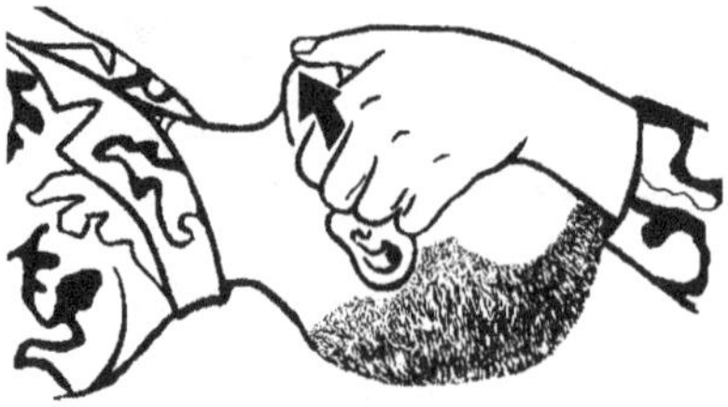

*Figure 2-4. Jaw-thrust technique of opening airway.*

(2) *Perform the head-tilt/chin-lift technique*. Place one hand on the casualty's forehead and apply firm, backward pressure with the palm to tilt the head back. Place the fingertips of the other hand under the bony part of the lower jaw and lift, bringing the chin forward. The thumb should not be used to lift the chin (Figure 2-5).

**NOTE**

The fingers should not press deeply into the soft tissue under the chin because the airway may be obstructed.

*Figure 2-5. Head-tilt/chin-lift technique of opening airway.*

(3) *Check for breathing (while maintaining an airway)*. After establishing an open airway, it is important to maintain that airway in an open position. Often the act of just opening and maintaining the airway will allow the casualty to breathe properly. Once the rescuer uses one of the techniques to open the airway (jaw-thrust or head-tilt/chin-lift), he should maintain that head position to keep the airway open. Failure to maintain the open airway will prevent the casualty from receiving an adequate supply of oxygen. Therefore, while maintaining an open airway the rescuer should check for breathing by observing the casualty's chest and performing the following actions within 3 to 5 seconds:

(*a*) **LOOK** for the chest to rise and fall.

(*b*) **LISTEN** for air escaping during exhalation by placing your ear near the casualty's mouth.

(*c*) **FEEL** for the flow of air on your cheek (see Figure 2-6).

(*d*) **PERFORM** rescue breathing if the casualty does not resume breathing spontaneously.

**NOTE**

If the casualty resumes breathing, monitor and maintain the open airway. He should be transported to an MTF, as soon as practical.

## 2-5. Rescue Breathing (Artificial Respiration)

a. If the casualty does not promptly resume adequate spontaneous breathing after the airway is open, rescue breathing (artificial respiration) must be started. Be calm! Think and act quickly! The sooner you begin rescue breathing, the more likely you are to restore the casualty's breathing. If you are in doubt whether the casualty is breathing, give artificial respiration, since it can do no harm to a person who is breathing. If the casualty is breathing, you can feel and see his chest move. If the casualty is breathing, you can feel and hear air being expelled by putting your hand or ear close to his mouth and nose.

*b*. There are several methods of administering rescue breathing. The mouth-to-mouth method is preferred; however, it cannot be used in all situations. If the casualty has a severe jaw fracture or mouth wound or his jaws are tightly closed by spasms, use the mouth-to-nose method.

## 2-6. Preliminary Steps—All Rescue Breathing Methods

*a*. Establish unresponsiveness. Call for help. Turn or position the casualty.

*b*. Open the airway.

*c*. Check for breathing by placing your ear over the casualty's mouth and nose, and looking toward his chest.

(1) **LOOK** for rise and fall of the casualty's chest (Figure 2-6).

(2) **LISTEN** for sounds of breathing.

(3) **FEEL** for breath on the side of your face. If the chest does not rise and fall and no air is exhaled, then the casualty is not breathing.

(4) **PERFORM** rescue breathing if the casualty is not breathing.

**NOTE**

Although the rescuer may notice that the casualty is making respiratory efforts, the airway may still be obstructed and opening the airway may be all that is needed. If the casualty resumes breathing, the rescuer should continue to maintain an open airway.

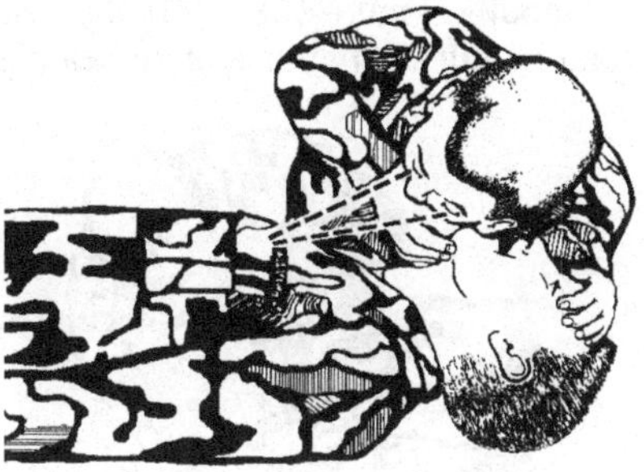

*Figure 2-6. Check for breathing.*

## 2-7. Mouth-to-Mouth Method

In this method of rescue breathing, you inflate the casualty's lungs with air from your lungs. This can be accomplished by blowing air into the person's mouth. The mouth-to-mouth rescue breathing method is performed as follows:

*a.* If the casualty is not breathing, place your hand on his forehead, and pinch his nostrils together with the thumb and index finger of this hand. Let this same hand exert pressure on his forehead to maintain the backward head tilt and maintain an open airway. With your other hand, keep your fingertips on the bony part of the lower jaw near the chin and lift (Figure 2-7).

*Figure 2-7. Head tilt/chin lift.*

**NOTE**

If you suspect the casualty has a neck injury and you are using the jaw-thrust technique, close the nostrils by placing your cheek tightly against them.

*b.* Take a deep breath and place your mouth (in an airtight seal) around the casualty's mouth (Figure 2-8). (If the injured person is small, cover both his nose and mouth with your mouth, sealing your lips against the skin of his face.)

*Figure 2-8. Rescue breathing.*

*c.* Blow two full breaths into the casualty's mouth (1 to 1 1/2 seconds per breath), taking a breath of fresh air each time before you blow. Watch out of the corner of your eye for the casualty's chest to rise. If the chest rises, sufficient air is getting into the casualty's lungs. Therefore, proceed as described in step (1). If the chest does not rise, do the following (*a*, *b*, and *c* below) and then attempt to ventilate again.

(1) Take corrective action immediately by reestablishing the airway. Make sure that air is not leaking from around your mouth or out of the casualty's pinched nose.

(2) Reattempt to ventilate.

(3) If the chest still does not rise, take the necessary action to open an obstructed airway (paragraph 2-10).

**NOTE**

If the initial attempt to ventilate the casualty is unsuccessful, reposition the casualty's head and repeat rescue breathing. Improper chin and head positioning is the most common cause of difficulty with ventilation. If the casualty cannot be ventilated after repositioning the head, proceed with foreign-body airway obstruction maneuvers (see paragraph 2-10).

(4) After giving two slow breaths, which cause the chest to rise, attempt to locate a pulse on the casualty. Feel for a pulse on the side of the casualty's neck closest to you by placing the first two fingers (index and middle fingers) of your hand on the groove beside the casualty's Adam's apple (carotid pulse) (Figure 2-9). (Your thumb should not be used for pulse taking because you may confuse your pulse beat with that of the casualty.) Maintain the airway by keeping your other hand on the casualty's forehead. Allow 5 to 10 seconds to determine if there is a pulse.

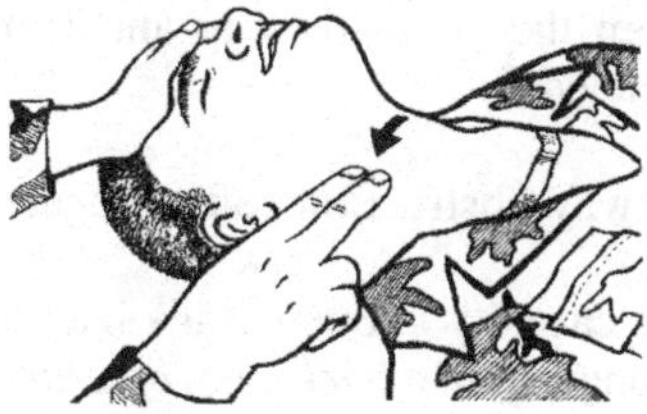

*Figure 2-9. Placement of fingers to detect pulse.*

(*a*) If signs of circulation are present and a pulse is found and the casualty is breathing—**STOP**; allow the casualty to breathe on his own. If possible, keep him warm and comfortable.

(*b*) If a pulse is found and the casualty is not breathing, continue rescue breathing.

(*c*) If a pulse is not found, seek medically trained personnel for help as soon as possible.

## 2-8. Mouth-to-Nose Method

Use this method if you cannot perform mouth-to-mouth rescue breathing because the casualty has a severe jaw fracture or mouth wound or his jaws are tightly closed by spasms. The mouth-to-nose method is performed in the same way as the mouth-to-mouth method except that you blow into the nose while you hold the lips closed with one hand at the chin. You then remove your mouth to allow the casualty to exhale passively. It may be necessary to separate the casualty's lips to allow the air to escape during exhalation.

## 2-9. Heartbeat

If a casualty's heart stops beating, you must immediately seek medical help. **SECONDS COUNT!** Stoppage of the heart is soon followed by cessation of respiration unless it has occurred first. Be calm! Think and act! When a casualty's heart has stopped, there is no pulse at all; the person is unconscious and limp, and the pupils of his eyes are open wide. When evaluating a casualty or when performing the preliminary steps of rescue breathing, feel for a pulse. If you DO NOT detect a pulse, seek medical help.

## 2-10. Airway Obstructions

In order for oxygen from the air to flow to and from the lungs, the upper airway must be unobstructed.

*a.* Upper airway obstructions often occur because—

(1) The casualty's tongue falls back into his throat while he is unconscious. The tongue *falls back* and *obstructs* the airway, it is not swallowed by the casualty.

**NOTE**

Ensure the correct positioning and maintenance of the open airway for an injured or unconscious casualty.

(2) Foreign bodies become lodged in the throat. These obstructions usually occur while eating. Choking on food (usually meat) is associated with—

- Attempting to swallow large pieces of poorly chewed food.
- Drinking alcohol.
- Slipping dentures.

(3) The contents of the stomach are regurgitated and may block the airway.

(4) Blood clots may form as a result of head and facial injuries.

*b.* Upper airway obstruction may cause either partial or complete airway blockage.

(1) *Partial airway obstruction.* The casualty may still have an air exchange. A good air exchange means that the casualty can cough

forcefully, though he may be wheezing between coughs. You, the rescuer, should not interfere, and should encourage the casualty to cough up the object obstructing his airway on his own. A poor air exchange may be indicated by weak coughing with a high pitched noise between coughs. Further, the casualty may show signs of shock (paragraph 1-6*b*[5]) indicating a need for oxygen. You should assist the casualty and treat him as though he had a complete obstruction.

(2) *Complete airway obstruction.* A complete obstruction (no air exchange) is indicated if the casualty cannot speak, breathe, or cough at all. He may be clutching his neck and moving erratically. In an unconscious casualty, a complete obstruction is also indicated if after opening his airway you cannot ventilate him.

## 2-11. Opening the Obstructed Airway—Conscious Casualty

Clearing a conscious casualty's airway obstruction can be performed with the casualty either standing or sitting and by following a relatively simple procedure.

> **WARNING**
>
> **Once an obstructed airway occurs, the brain will develop an oxygen deficiency resulting in unconsciousness. Death will follow rapidly if breathing is not promptly restored.**

*a.* Ask the casualty if he can speak or if he is choking. Check for the universal choking sign (Figure 2-10).

*Figure 2-10. Universal sign of choking.*

*b.* If the casualty can speak, encourage him to attempt to cough; the casualty still has a good air exchange. If he is able to speak or cough effectively, DO NOT interfere with his attempts to expel the obstruction.

*c.* Listen for high pitched sounds when the casualty breathes or coughs (poor air exchange). If there is poor air exchange or no breathing, CALL FOR HELP and immediately deliver manual thrusts (either an abdominal or chest thrust).

**NOTE**

The manual thrust with the hands centered between the waist and the rib cage is called an abdominal thrust (or Heimlich maneuver). The chest thrust (the hands are centered in the middle of the breastbone) is used only for an individual in the advanced stages of pregnancy, in the markedly obese casualty, or if there is a significant abdominal wound.

(1) Apply abdominal thrusts. This can be accomplished by using the following procedures:

(*a*) Stand behind the casualty and wrap your arms around his waist.

(*b*) Make a fist with one hand and grasp it with the other. The thumb side of your fist should be against the casualty's abdomen, in the midline and slightly above the casualty's navel, but well below the tip of the breastbone (Figure 2-11).

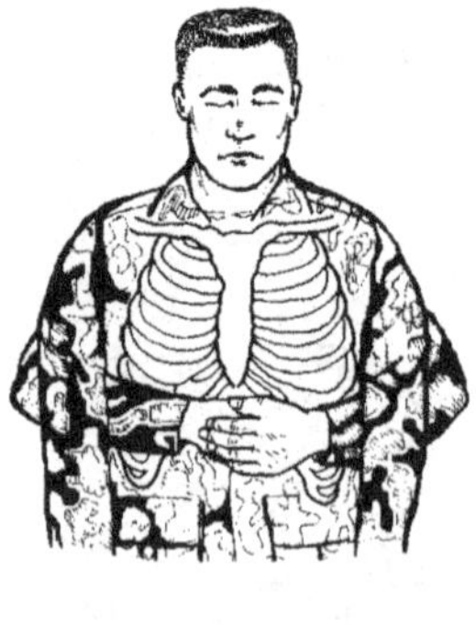

*Figure 2-11. Anatomical view of abdominal thrust procedure.*

(*c*) Press the fists into the abdomen with a quick backward and upward thrust (Figure 2-12).

*Figure 2-12. Profile view of abdominal thrust.*

(*d*) Each thrust should be a separate and distinct movement.

**NOTE**

Continue performing abdominal thrusts until the obstruction is expelled or the casualty becomes unresponsive.

(*e*) If the casualty becomes unresponsive, call for help as you proceed with steps to open the airway, and perform rescue breathing. (Refer to paragraph 2-7 for information on how to perform mouth-to-mouth resuscitation.)

(2) Apply chest thrusts. An alternate technique to the abdominal thrust is the chest thrust. This technique is useful when the casualty has an abdominal wound, when the casualty is pregnant, or when the casualty is so large that you cannot wrap your arms around the abdomen. To apply chest thrusts with casualty sitting or standing:

(*a*) Stand behind the casualty and wrap your arms around his chest with your arms under his armpits.

(*b*) Make a fist with one hand and place the thumb side of the fist in the middle of the breastbone (take care to avoid the tip of the breastbone and the margins of the ribs).

(*c*) Grasp the fist with the other hand and exert thrusts (Figure 2-13).

*Figure 2-13. Profile view of chest thrust.*

(*d*) Each thrust should be delivered slowly, distinctly, and with the intent of relieving the obstruction.

(*e*) Perform chest thrusts until the obstruction is expelled or the casualty becomes unresponsive.

(*f*) If the casualty becomes unresponsive, call for help as you proceed with steps to open the airway and perform rescue breathing.

## 2-12. Opening the Obstructed Airway—Casualty Lying Down or Unresponsive

The following procedures are used to expel an airway obstruction in a casualty who is lying down, who becomes unconscious, or who is found unconscious (the cause unknown):

- If a conscious casualty who is choking becomes unresponsive, call for help, open the airway, perform a finger sweep, and attempt rescue breathing (paragraphs 2-4 through 2-8). If you still cannot administer rescue breathing due to an airway blockage, then remove the airway obstruction using the procedures as in *b* below.

- If a casualty is unresponsive when you find him (the cause unknown), assess or evaluate the situation, call for help, position the casualty on his back, open the airway, establish breathlessness, and attempt to perform rescue breathing (paragraphs 2-4 through 2-8).

*a.* Open the airway and attempt rescue breathing (refer to paragraph 2-7 for information on how to perform mouth-to-mouth resuscitation).

*b.* If still unable to ventilate the casualty, perform 6 to 10 manual (abdominal or chest) thrusts.

(1) To perform the abdominal thrusts:

(*a*) Kneel astride the casualty's thighs (Figure 2-14).

*Figure 2-14. Abdominal thrust on unresponsive casualty.*

(*b*) Place the heel of one hand against the casualty's abdomen (in the midline slightly above the navel but well below the tip of the breastbone). Place your other hand on top of the first one. Point your fingers toward the casualty's head.

(*c*) Press into the casualty's abdomen with a quick, forward and upward thrust. You can use your body weight to perform the maneuver. Deliver each thrust quickly and distinctly.

(*d*) Repeat the sequence of abdominal thrusts, finger sweep, and rescue breathing (attempt to ventilate) as long as necessary to remove the object from the obstructed airway.

(*e*) If the casualty's chest rises, proceed to feeling for pulse.

(2) To perform chest thrusts:

(*a*) Place the unresponsive casualty on his back, face up, and open his mouth. Kneel close to the side of the casualty's body.

*1.* Locate the lower edge of the casualty's ribs with your fingers. Run the fingers up along the rib cage to the notch (Figure 2-15A).

*2.* Place the middle finger on the notch and the index finger next to the middle finger on the lower edge of the breastbone. Place the heel of the other hand on the lower half of the breastbone next to the two fingers (Figure 2-15B).

*3.* Remove the fingers from the notch and place that hand on top of the positioned hand on the breastbone, extending or interlocking the fingers (Figure 2-15C).

*4.* Straighten and lock your elbows with your shoulders directly above your hands without bending the elbows, rocking, or allowing the shoulders to sag. Apply enough pressure to depress the breastbone 1 1/2 to 2 inches, then release the pressure completely (Figure 2-15D). Do this 6 to 10 times. Each thrust should be delivered quickly and distinctly. See Figure 2-16 for another view of the breastbone being depressed.

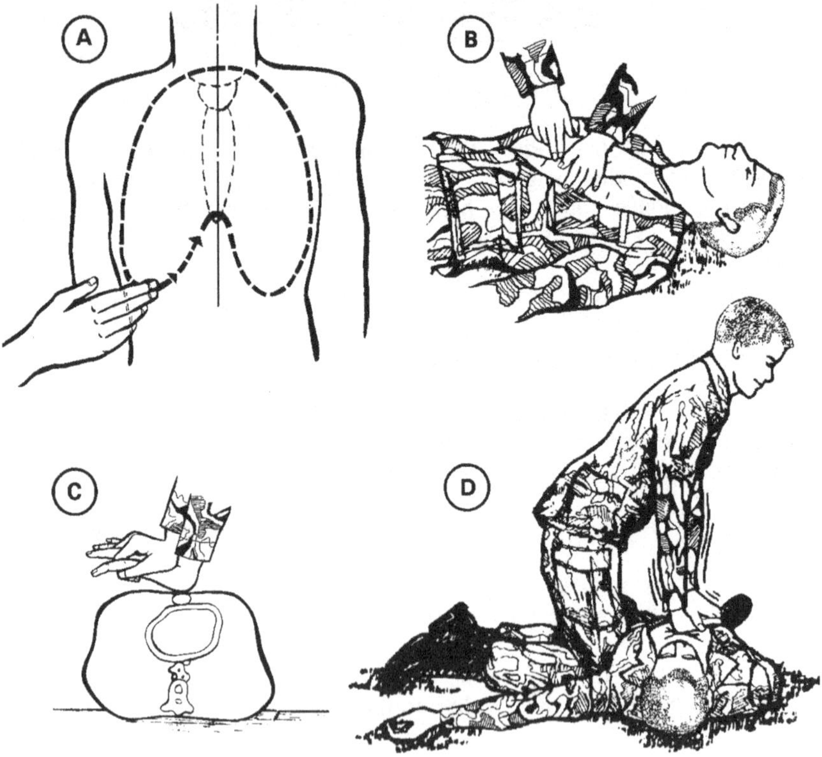


*Figure 2-15. Hand placement for chest thrust (Illustrated A-D).*

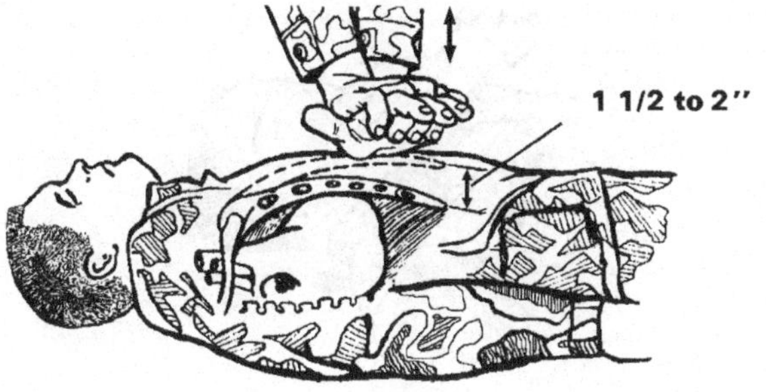


*Figure 2-16. Breastbone depressed 1 1/2 to 2 inches.*

(*b*) Repeat the sequence of chest thrust, finger sweep, and rescue breathing as long as necessary to clear the object from the obstructed airway. See paragraph (3) below.

(*c*) If the casualty's chest rises, proceed to feeling for his pulse.

(3) If you still cannot administer rescue breathing due to an airway obstruction, then remove the airway obstruction using the procedures in steps (*a*) and (*b*) below.

(*a*) Place the casualty on his back, face up, turn the unresponsive casualty as a unit, and call out for help.

(*b*) Perform finger sweep, keep casualty face up, use tongue-jaw lift to open mouth.

*1.* Open the casualty's mouth by grasping both his tongue and lower jaw between your thumb and fingers and lifting (tongue-jaw lift) (Figure 2-17). If you are unable to open his mouth, cross your fingers and thumb (crossed-finger method) and push his teeth apart (Figure 2-18) by pressing your thumb against his upper teeth and pressing your finger against his lower teeth.

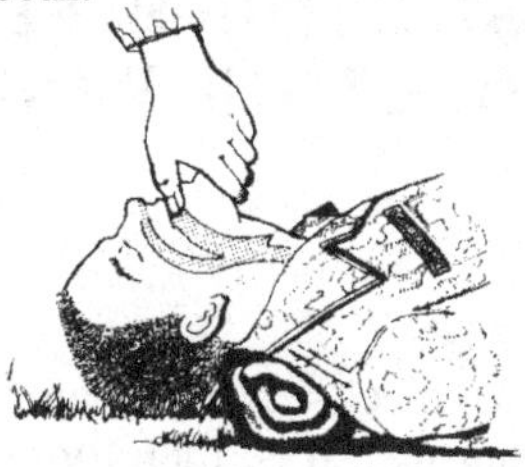

*Figure 2-17. Opening casualty's mouth (tongue-jaw lift).*

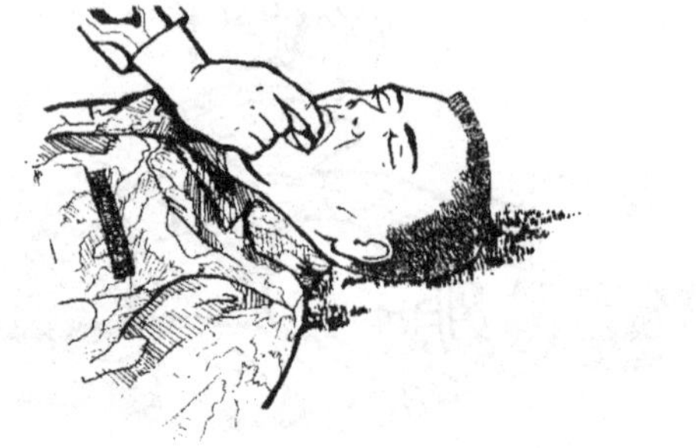

*Figure 2-18. Opening casualty's mouth (crossed-finger method).*

2. Insert the index finger of the other hand down along the inside of his cheek to the base of the tongue. Use a hooking motion from the side of the mouth toward the center to dislodge the foreign body (Figure 2-19).

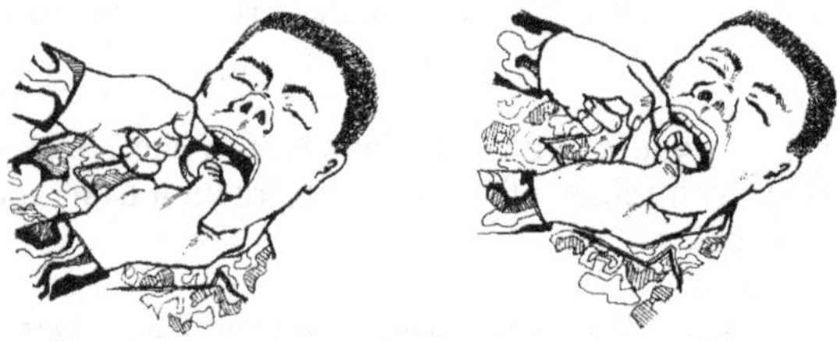

*Figure 2-19. Using finger to dislodge a foreign body.*

**WARNING**

**Take care not to force the object deeper into the airway by pushing it with the finger.**

# Section II. STOP THE BLEEDING AND PROTECT THE WOUND

## 2-13. General

The longer a service member bleeds from a major wound, the less likely he will be able to survive his injuries. It is, therefore, important that the first aid provider promptly stop the external bleeding.

## 2-14. Clothing

In evaluating the casualty for location, type, and size of the wound or injury, cut or tear his clothing and carefully expose the entire area of the wound. This procedure is necessary to properly visualize injury and avoid further contamination. Clothing stuck to the wound should be left in place to avoid further injury. DO NOT touch the wound; keep it as clean as possible.

**WARNING**

**DO NOT REMOVE protective clothing in a chemical environment. Apply dressings over the protective clothing.**

## 2-15. Entrance and Exit Wounds

Before applying the dressing, carefully examine the casualty to determine if there is more than one wound. A missile may have entered at one point and exited at another point. The *EXIT* wound is usually *LARGER* than the entrance wound.

**WARNING**

**The casualty should be continually monitored for development of conditions which may require the performance of necessary basic lifesaving measures, such as clearing the airway and mouth-to-mouth resuscitation. All open (or penetrating) wounds should be checked for a point of entry and exit and first aid measures applied accordingly.**

**WARNING**

**If the missile lodges in the body (fails to exit), DO NOT attempt to remove it or probe the wound. Apply a dressing. If there is an object extending from (impaled in) the wound, DO NOT remove the object. Apply a dressing around the object and use additional improvised bulky materials/dressings (use the cleanest material available) to build up the area around the object to stabilize the object and prevent further injury. Apply a supporting bandage over the bulky materials to hold them in place.**

## 2-16. Field Dressing

*a.* Use the casualty's field dressing; remove it from the wrapper and grasp the tails of the dressing with both hands (Figure 2-20).

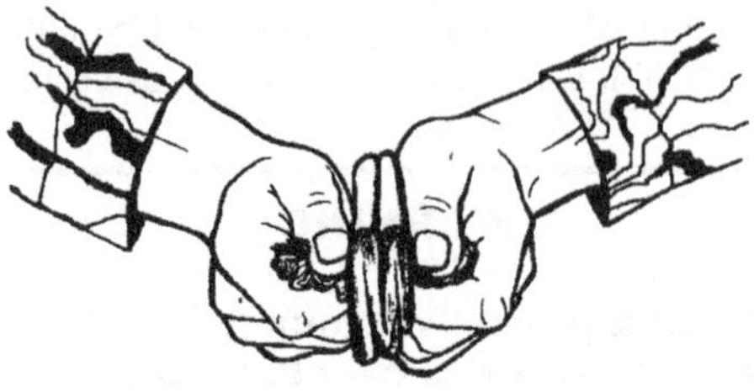

*Figure 2-20. Grasping tails of dressing with both hands.*

> **WARNING**
>
> **DO NOT touch the white (sterile) side of the dressing, and DO NOT allow it to come in contact with any surface other than the wound.**

*b.* Hold the dressing directly over the wound with the white side down. Pull the dressing open (Figure 2-21) and place it directly over the wound (Figure 2-22).

*Figure 2-21. Pulling dressing open.*

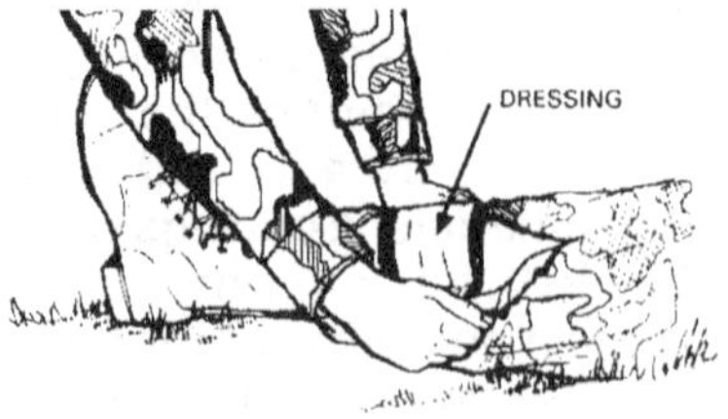


*Figure 2-22. Placing dressing directly on wound.*

*c.* Hold the dressing in place with one hand. Use the other hand to wrap one of the tails around the injured part, covering about one-half of the dressing (Figure 2-23). Leave enough of the tail for a knot. If the casualty is able, he may assist by holding the dressing in place.

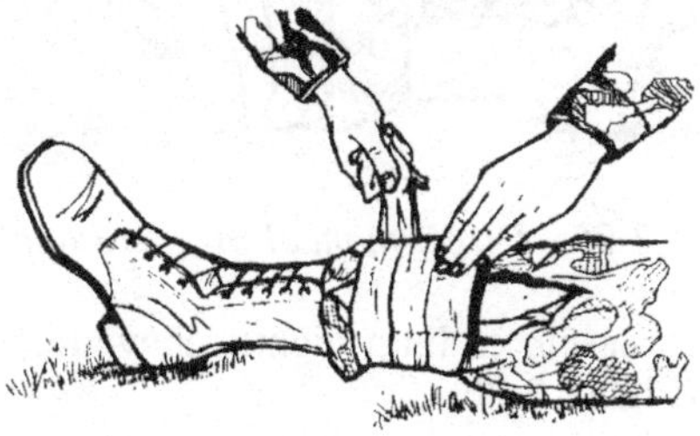

*Figure 2-23. Wrapping tail of dressing around injured part.*

*d.* Wrap the other tail in the opposite direction until the remainder of the dressing is covered. The tails should seal the sides of the dressing to keep foreign material from getting under it.

*e.* Tie the tails into a nonslip knot over the outer edge of the dressing (Figure 2-24). **DO NOT TIE THE KNOT OVER THE WOUND.** In order to allow blood to flow to the rest of an injured limb, tie the dressing firmly enough to prevent it from slipping but without causing a tourniquet-like effect; that is, the skin beyond the injury should not becomes cool, blue, or numb.

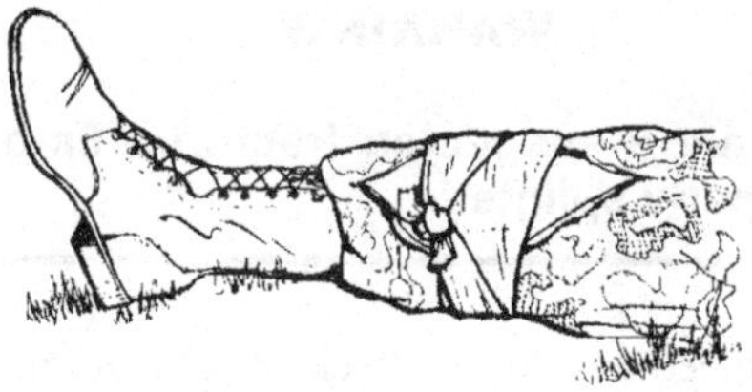

*Figure 2-24. Tails tied into nonslip knot.*

## 2-17. Manual Pressure

*a.* If bleeding continues after applying the sterile field dressing, direct manual pressure may be used to help control bleeding. Apply such pressure by placing a hand on the dressing and exerting firm pressure for 5 to 10 minutes (Figure 2-25). The casualty may be asked to do this himself if he is conscious and can follow instructions.

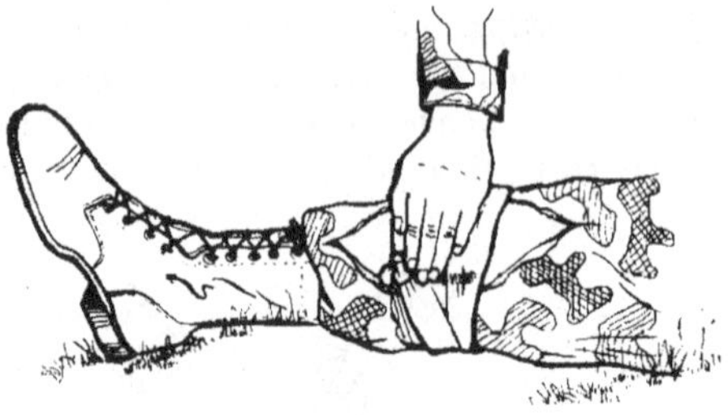

*Figure 2-25. Direct manual pressure applied.*

*b.* Elevate an injured limb slightly above the level of the heart to reduce the bleeding (Figure 2-26).

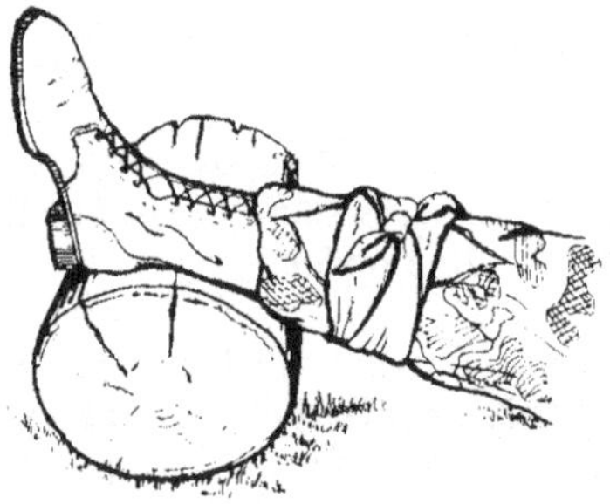

*Figure 2-26. Injured limb elevated.*

**WARNING**

**DO NOT elevate a suspected fractured limb unless it has been properly splinted.**

*c.* If the bleeding stops, check shock; administer first aid for shock as necessary. If the bleeding continues, apply a pressure dressing.

### 2-18. Pressure Dressing

Pressure dressings aid in blood clotting and compress the open blood vessel. If bleeding continues after the application of a field dressing, manual pressure, and elevation, then a pressure dressing must be applied as follows:

*a.* Place a wad of padding on top of the field dressing, directly over the wound (Figure 2-27). Keep the injured extremity elevated.

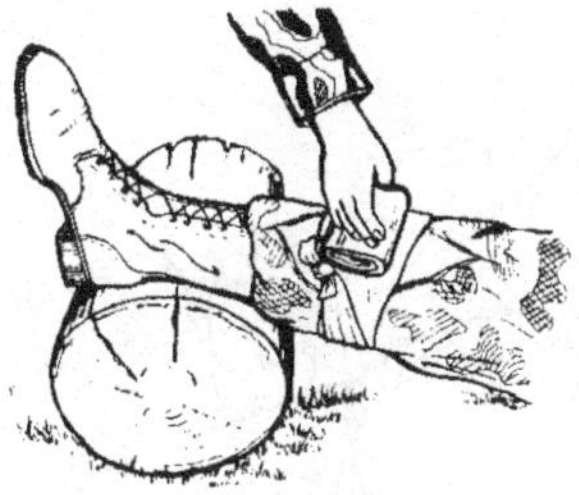

*Figure 2-27. Wad of padding on top of field dressing.*

**NOTE**

Improvised bandages may be made from strips of cloth. These strips may be made from T-shirts, socks, or other garments.

*b.* Place an improvised dressing (or cravat, if available) over the wad of padding (Figure 2-28). Wrap the ends tightly around the injured limb, covering the previously placed field dressing (Figure 2-29).

*Figure 2-28. Improvised dressing over wad of padding*

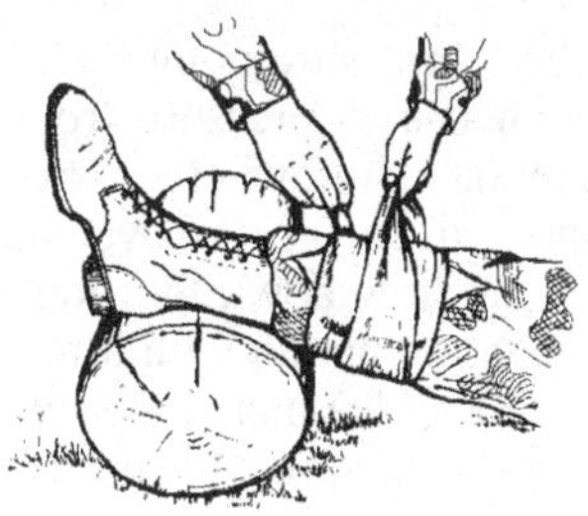

*Figure 2-29. Ends of improvised dressing wrapped tightly around limb.*

*c.* Tie the ends together in a nonslip knot, directly over the wound site (Figure 2-30). DO NOT tie so tightly that it has a tourniquet-like effect. If bleeding continues and all other measures have failed, or if the limb is severed, then apply a tourniquet. Use the tourniquet as a **LAST RESORT**. When the bleeding stops, check for shock; administer first aid for shock as necessary.

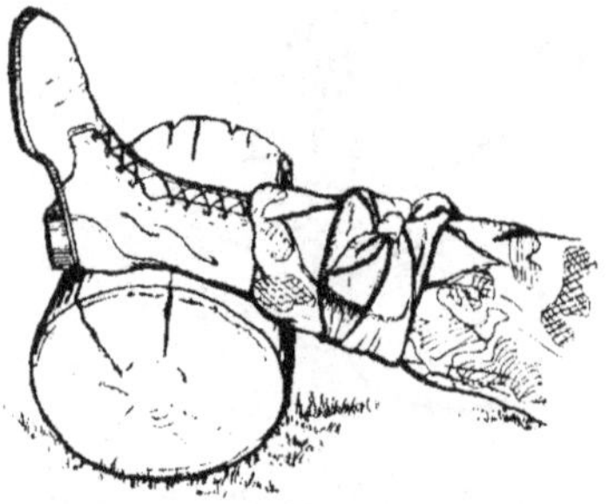

*Figure 2-30. Ends of improvised dressing tied together in nonslip knot.*

**NOTE**

Distal end of wounded extremities (fingers and toes) should be checked periodically for adequate circulation. The dressing must be loosened if the extremity becomes cool, blue, or numb.

**NOTE**

If bleeding continues and all other measures have failed (dressings and covering wound, applying direct manual pressure, elevating the limb above the heart level, and applying a pressure dressing while maintaining limb elevation) *then apply digital pressure* (see paragraph 2-19).

## 2-19. Digital Pressure

Digital pressure (often called "pressure points") is an alternative method to control bleeding. This method uses pressure from the fingers, thumbs, or hands to press at the site or point where a main artery supplying the wounded area lies near the skin surface or over bone (Figure 2-31). This pressure may help shut off or slow down the flow of blood from the heart to the wound and is used in combination with direct pressure and elevation. It may help in instances where bleeding is not easily controlled, where a pressure dressing has not yet been applied, or where pressure dressings are not readily available.

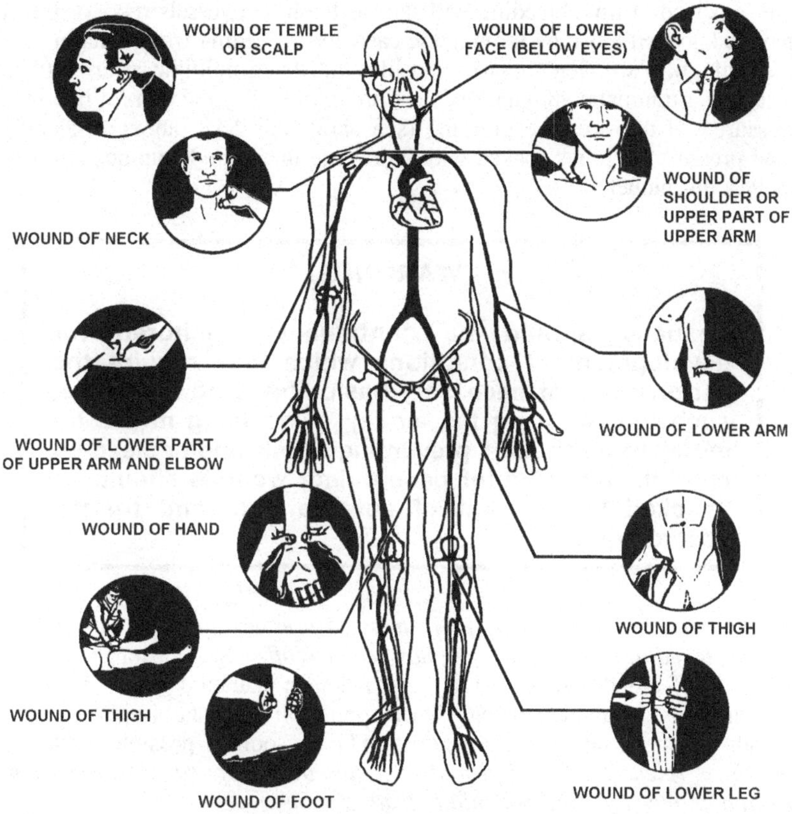


*Figure 2-31. Digital pressure (pressure with fingers, thumbs or hands).*

## 2-20. Tourniquet

**DANGER**

**A tourniquet is only used on an arm or leg where there is a danger of the casualty losing his life (bleeding to death).**

A tourniquet is a constricting band placed around an arm or leg to control bleeding. A service member whose arm or leg has been completely amputated may not be bleeding when first discovered, but a tourniquet should be applied anyway. This absence of bleeding is due to the body's normal defenses (contraction or clotting of blood vessels) as a result of the amputation, but after a period of time bleeding will start as the blood vessels relax or the clot may be knocked loose by moving the casualty. Bleeding from a major artery of the thigh, lower leg, or arm and bleeding from multiple arteries (which occurs in a traumatic amputation) may prove to be beyond control by manual pressure. If the pressure dressing (see paragraph 2-18, above) under firm hand pressure becomes soaked with blood and the wound continues to bleed, apply a tourniquet.

**WARNING**

**Casualty should be continually monitored for development of conditions which may require the performance of necessary basic lifesaving measures, such as: clearing the airway, performing mouth-to-mouth resuscitation, preventing shock, and/or bleeding control. All open (or penetrating) wounds should be checked for a point of entry or exit and treated accordingly.**

*The tourniquet should not be used unless a pressure dressing has failed to stop the bleeding or an arm or leg has been cut off.* On occasion, tourniquets have injured blood vessels and nerves. If left in place too long, a tourniquet can cause loss of an arm or leg. Once applied, it must stay in place, and the casualty must be taken to the nearest MTF as soon as possible. *DO NOT loosen or release a tourniquet after it has been applied as release could precipitate bleeding and potentially lead to shock.*

*a. Improvising a Tourniquet.* In the absence of a specially designed tourniquet, a tourniquet may be made from a strong, pliable material, such as gauze or muslin bandages, clothing, or cravats. An improvised tourniquet is used with a rigid stick-like object. To minimize skin damage, ensure that the improvised tourniquet is at least 2 inches wide.

**WARNING**

**The tourniquet must be easily identified or easily seen.**

**WARNING**

**DO NOT use wire or shoestring for a tourniquet band.**

*b. Placing the Improvised Tourniquet.*

(1) Place the tourniquet around the limb, between the wound and the body trunk (or between the wound and the heart). Never place it directly over a wound, a fracture, or joint. Tourniquets, for maximum effectiveness, should be placed on the upper arm or above the knee on the thigh (Figure 2-32).

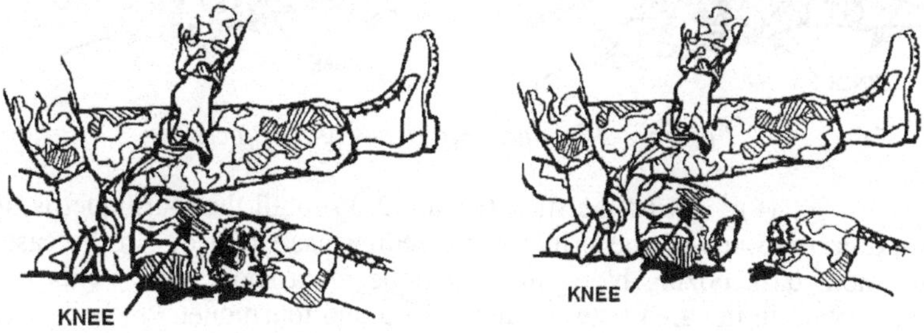


*Figure 2-32. Tourniquet above knee.*

(2) The tourniquet should be well-padded. If possible, place the tourniquet over the smoothed sleeve or trouser leg to prevent the skin from being pinched or twisted. If the tourniquet is long enough, wrap it around the limb several times, keeping the material as flat as possible. Damaging the skin may deprive the surgeon of skin required to cover an amputation. Protection of the skin also reduces pain.

*c. Applying the Tourniquet.*

(1) Tie a half-knot. (A half-knot is the same as the first part of tying a shoe lace.)

(2) Place a stick (or similar rigid object) on top of the half-knot (Figure 2-33).

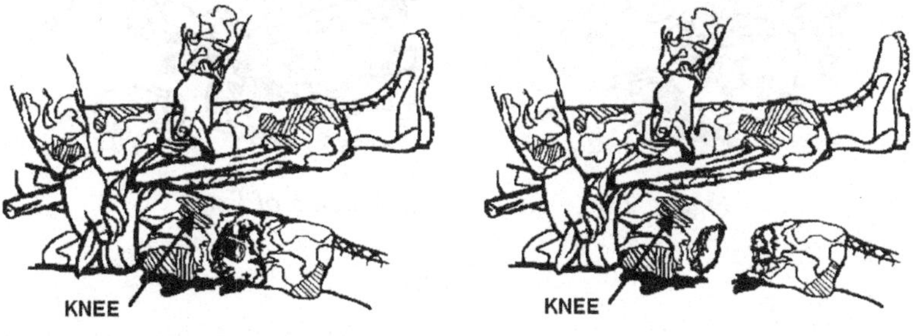


*Figure 2-33. Rigid object on top of half-knot.*

(3) Tie a full knot over the stick (Figure 2-34).

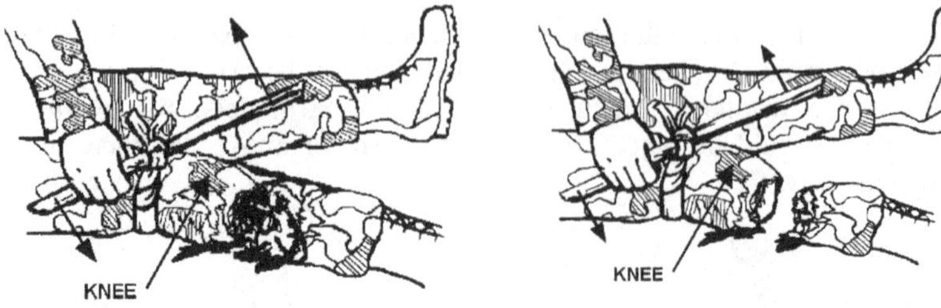


*Figure 2-34. Full knot over rigid object.*

(4) Twist the stick (Figure 2-35) until the tourniquet is tight around the limb and/or the bright red bleeding has stopped. In the case of amputation, dark oozing blood may continue for a short time. This is the blood trapped in the area between the wound and tourniquet.

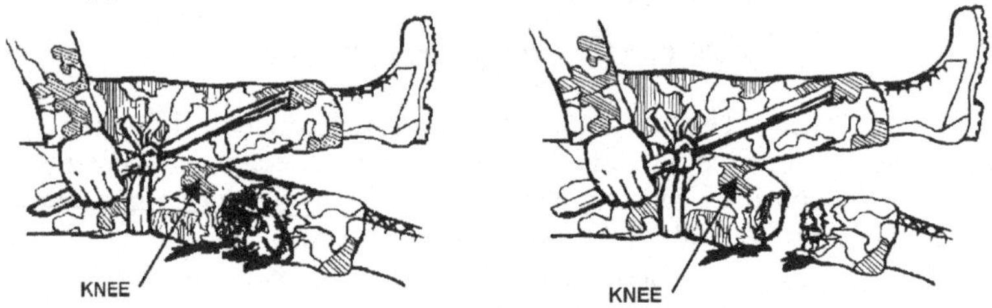


*Figure 2-35. Stick twisted.*

(5) Fasten the tourniquet to the limb by looping the free ends of the tourniquet over the ends of the stick. Then bring the ends around the limb to prevent the stick from loosening. Tie them together on the side of the limb (Figure 2-36).

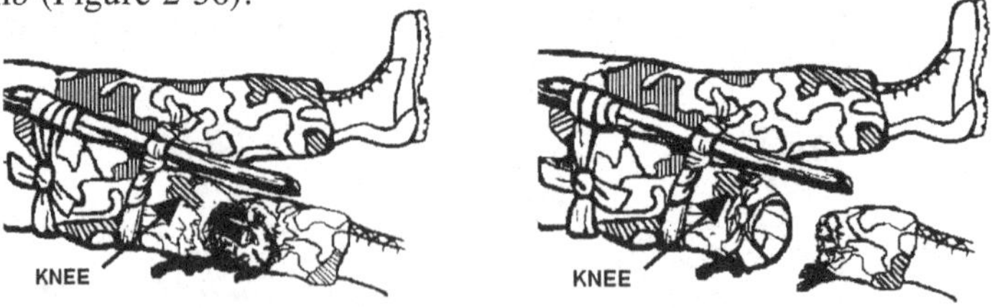


*Figure 2-36. Tie free ends on side of limb.*

**NOTE**

Other methods of securing the stick may be used as long as the stick does not unwind and no further injury results.

**NOTE**

If possible, save and transport any severed (amputated) limbs or body parts with (but out of sight of) the casualty.

(6) DO NOT cover the tourniquet—you should leave it in full view. If the limb is missing (total amputation), apply a dressing to the stump. All wounds should have a dressing to protect the wound from contamination.

(7) Mark the casualty's forehead with a "T" and the time to indicate a tourniquet has been applied. If necessary, use the casualty's blood to make this mark.

(8) Check and treat for shock.

(9) Seek medical aid.

**CAUTION**

Only appropriately skilled medical personnel may adjust or otherwise remove/release the tourniquet in the appropriate setting.

# Section III. CHECK FOR SHOCK AND ADMINISTER FIRST AID MEASURES

## 2-21. General

The term *shock* has a variety of meanings. In medicine, it refers to a collapse of the body's cardiovascular system which includes an inadequate supply of blood to the body's tissues. Shock stuns and weakens the body. When the normal blood flow in the body is upset, death can result. Early recognition and proper first aid may save the casualty's life.

## 2-22. Causes and Effects

*a.* There are three basic mechanisms associated with shock. These are—

- The heart is damaged and fails to work as a pump.

- Blood loss (heavy bleeding) causes the volume of fluid within the vascular system to be insufficient.

- The blood vessels dilate (open wider) so that the blood within the system (even though it is a normal volume [the casualty is not bleeding or dehydrated]) is insufficient to provide adequate circulation within the body.

*b.* Shock may be the result of a number of conditions. These include—

- Dehydration.

- Allergic reaction to foods, drugs, insect stings, and snakebites.

- Significant loss of blood.

- Reaction to the sight of a wound, blood, or other traumatic scene.

- Traumatic injuries, such as—

  - Burns.

  - Gunshot or shrapnel wounds.

  - Crush injuries.

  - Blows to the body (which can cause broken bones or damage to internal organs).

  - Head injuries.

  - Penetrating wounds (such as from a knife, bayonet, or missile).

## 2-23. Signs and Symptoms of Shock

Examine the casualty to see if he has any of the following signs and symptoms:

- Sweaty but cool skin (clammy skin).

- Weak and rapid pulse.
- Paleness of skin (in dark-skinned individuals they may have a grayish look to their skin).
- Restlessness, nervousness.
- Thirst.
- Loss of blood (bleeding).
- Confusion (or loss of awareness).
- Faster-than-normal breathing rate.
- Blotchy or bluish skin (especially around the mouth and lips).
- Nausea and/or vomiting.

## 2-24. First Aid Measures for Shock

In the field, the first aid procedures administered for shock are identical to procedures that would be performed *to prevent shock.* When treating a casualty, assume that shock is present or will occur shortly. By waiting until actual signs and symptoms of shock are noticeable, the rescuer may jeopardize the casualty's life.

*a. Position the Casualty.* (DO NOT move the casualty or his limbs if suspected fractures have not been splinted. See Chapter 4 for details.)

(1) Move the casualty to cover, if cover is available and the situation permits.

(2) Lay the casualty on his back.

**NOTE**

A casualty in shock from a chest wound or one who is experiencing breathing difficulty, may breathe easier in a sitting position. If this is the case, allow him to sit upright, but monitor carefully in case his condition worsens.

(3) Elevate the casualty's feet higher than the level of his heart. Use a stable object (field pack or rolled up clothing) so that his feet will not slip off (Figure 2-37).

**WARNING**

**DO NOT elevate legs if the casualty has an unsplinted broken leg, head injury, or abdominal injury.**

*Figure 2-37. Clothing loosened and feet elevated.*

**WARNING**

**Check casualty for leg fracture(s) and splint, if necessary, before elevating his feet. For a casualty with an abdominal wound, place his knees in an upright (flexed) position.**

(4) Loosen clothing at the neck, waist, or wherever it may be binding.

**CAUTION**

DO NOT loosen or remove protective clothing in a chemical environment.

(5) Prevent chilling or overheating. The key is to maintain body temperature. In cold weather, place a blanket or other like item over him to keep him warm and under him to prevent chilling (Figure 2-38). However, if a tourniquet has been applied, leave it exposed (if possible). In hot weather, place the casualty in the shade and protect him from becoming chilled; however, avoid the excessive use of blankets or other coverings.

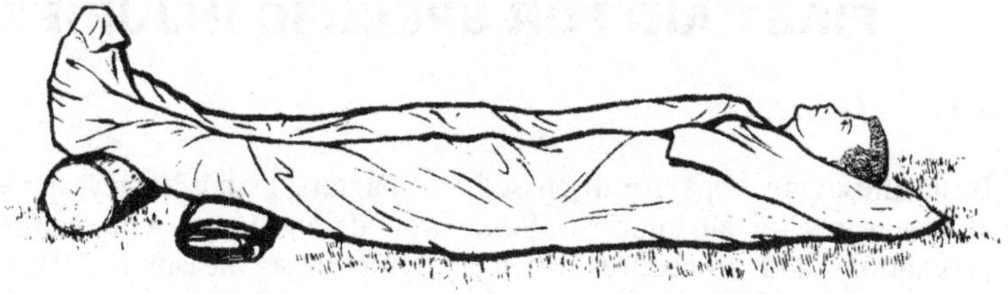

*Figure 2-38. Body temperature maintained.*

(6) Calm the casualty. Throughout the entire procedure of providing first aid for a casualty, the rescuer should reassure the casualty and keep him calm. This can be done by being authoritative (taking charge) and by showing self-confidence. Assure the casualty that you are there to help him.

(7) Seek medical aid.

*b. Food and/or Drink.* When providing first aid for shock, DO NOT give the casualty any food or drink. If you must leave the casualty or if he is unconscious, turn his head to the side to prevent him from choking if he vomits (Figure 2-39).

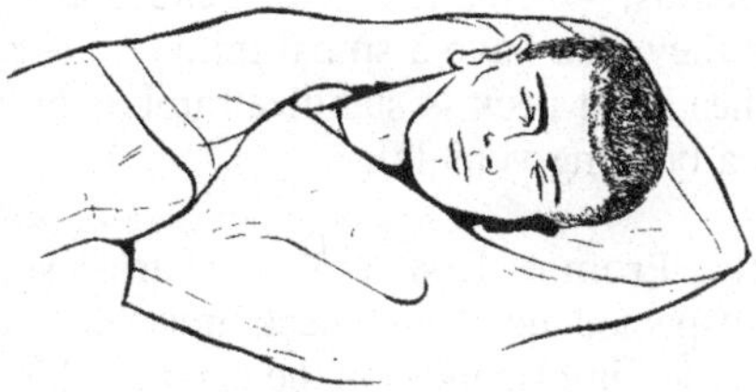

*Figure 2-39. Casualty's head turned to side.*

*c. Evaluate Casualty.* Continue to evaluate the casualty until medical personnel arrives or the casualty is transported to an MTF.

CHAPTER 3

# FIRST AID FOR SPECIFIC INJURIES

## 3-1. General

Basic lifesaving steps are discussed in Chapters 1 and 2; they apply to first aid measures for all injuries. Some wounds and burns will require special precautions and procedures when applying these measures. This chapter discusses specific first aid procedures for wounds of the head, face, and neck; chest and stomach wounds; and burns. It also discusses the techniques for applying dressings and bandages to specific parts of the body.

## 3-2. Head, Neck, and Facial Injuries

*a. Head Injuries.*

(1) Head injuries range from minor abrasions or cuts on the scalp to severe brain injuries that may result in unconsciousness and sometimes death. Head injuries are classified as open or closed wounds. An open wound is one that is visible, has a break in the skin, and usually has evidence of bleeding. A closed wound may be visible (such as a depression in the skull) or the first aid provider may not be able to see any apparent injury (such as internal bleeding). Some head injuries result in unconsciousness; however, a service member may have a serious head wound and still be conscious. Casualties with head and neck injuries should be treated as though they also have a spinal injury. The casualty should not be moved until the head and neck is stabilized unless he is in immediate danger (such as close to a burning vehicle).

(2) Prompt first aid measures should be initiated for casualties with suspected head and neck injuries. The conscious casualty may be able to provide information on the extent of his injuries. However, as a result of the head injury, he may be confused and unable to provide accurate information. The signs and symptoms a first aid provider might observe are—

- Nausea and vomiting.
- Convulsions or twitches.
- Slurred speech.
- Confusion and loss of memory. (Does he know who he is? Does he know where he is? Does he know what day it is?)
- Recent unconsciousness.

- Dizziness.
- Drowsiness.
- Blurred vision, unequal pupils, or bruising (black eyes).
- Paralysis (partial or full).
- Complaint of headache.
- Bleeding or other fluid discharge from the scalp, nose, or ears.
- Deformity of the head (depression or swelling).
- Staggering while walking.

*b. Neck Injuries.* Neck injuries may result in heavy bleeding. Apply pressure above and below the injury, *but do not interfere with the breathing process*, and attempt to control the bleeding. Apply a dressing. Always evaluate the casualty for a possible neck fracture/spinal cord injury; if suspected, seek medical treatment immediately.

**NOTE**

Establish and maintain the airway in cases of facial or neck injuries. If a neck fracture or spinal cord injury is suspected, immobilize the injury and, if necessary, perform basic life support measures.

*c. Facial Injuries.* Soft tissue injuries of the face and scalp are common. Abrasions (scrapes) of the skin cause no serious problems. Contusions (injury without a break in the skin) usually cause swelling. A contusion of the scalp looks and feels like a lump. Laceration (cut) and avulsion (torn away tissue) injuries are also common. Avulsions are frequently caused when a sharp blow separates the scalp from the skull beneath it. Because the face and scalp are richly supplied with blood vessels (arteries and veins), wounds of these areas usually bleed heavily.

## 3-3. General First Aid Measures

*a. General Considerations.* The casualty with a head injury (or suspected head injury) should be continually monitored for the development of conditions that *may require* basic lifesaving measures. After initiating first aid measures, request medical assistance and evacuation. If dedicated medical evacuation assets are not available, transport the casualty to an MTF as soon as the situation permits. The first aid provider should not attempt to remove a protruding object from the head or give the casualty anything to eat or drink. Further, the first aid provider should be prepared to—

- Clear the airway.
- Control bleeding (external).
- Administer first aid measures for shock.
- Keep the casualty warm.
- Protect the wound.

*b. Unconscious Casualty*. An unconscious casualty does not have control of all of his body's functions and may choke on his tongue, blood, vomitus, or other substances. (Refer to Figure 2-39.)

(1) *Breathing*. The brain requires a constant supply of oxygen. A bluish (or in an individual with dark skin—grayish) color of skin around the lips and nail beds indicates that the casualty is not receiving enough oxygen. Immediate action must be taken to clear the airway, to position the casualty on his side, or to initiate rescue breathing.

(2) *Bleeding*. Bleeding from a head injury usually comes from blood vessels within the scalp. Bleeding can also develop inside the skull or within the brain. In most instances visible bleeding from the head can be controlled by application of the field first aid dressing.

**CAUTION**

**DO NOT** attempt to put unnecessary pressure on the wound or attempt to push any brain matter back into the head (skull). **DO NOT** apply a pressure dressing.

*c. Concussion*. If an individual receives a heavy blow to the head or face, he may suffer a brain concussion (an injury to the brain that involves a temporary loss of some or all of the brain's ability to function). For example, the casualty may not breathe properly for a short period of time, or he may become confused and stagger when he attempts to walk. Symptoms of a concussion may only last for a short period of time. However, if a casualty is suspected of having suffered a concussion, he should be transported to an MTF as soon as conditions permit.

*d. Convulsions*. Convulsions (seizures/involuntary jerking) may occur even after a mild head injury. When a casualty is convulsing, protect him from hurting himself. Take the following measures:

(1) Ease him to the ground if he is standing or sitting.

(2) Support his head and neck.

(3) Maintain his airway.

(4) Protect him from further injury (such as hitting close-by objects).

**NOTE**

**DO NOT** forcefully hold the arms and legs if they are jerking because this can lead to broken bones. **DO NOT** force anything between the casualty's teeth—especially if they are tightly clenched because this may obstruct the casualty's airway. Maintain the casualty's airway if necessary.

*e. Brain Damage*. In *severe* head injuries where brain tissue is protruding, *leave the wound alone*; carefully place a loose moistened dressing (moistened with sterile normal saline if available) and also a first aid dressing over the tissue to protect it from further contamination. ***DO NOT** remove or disturb any foreign matter that may be in the wound*. Position the casualty so that his head is higher than his body. Keep him warm and *seek medical assistance immediately*.

**NOTE**

If there is an object extending from the wound, **DO NOT** remove the object. Improvise bulky dressings from the cleanest material available and place this material around the protruding object for support, then apply the field dressing.

## 3-4. Chest Wounds

Blunt trauma, bullet or missile wounds, stab wounds, or falls may cause chest injuries. These injuries can be serious and may cause death quickly if first aid is not administered in a timely manner. A casualty with a chest injury may complain of pain in the chest or shoulder area; he may have difficulty breathing. His chest may not rise normally when he breathes. The injury may cause the casualty to cough up blood and to have a rapid or a weak heartbeat. A casualty with an open chest wound has a punctured chest wall. The sucking sound heard when he breathes is caused by air leaking into his chest cavity. This particular type of wound is dangerous and will collapse the injured lung (Figure 3-1). Breathing becomes difficult for the casualty because the wound is open. The service members life may depend upon how quickly you apply an occlusive dressing over the wound (refer to paragraph 3-5).

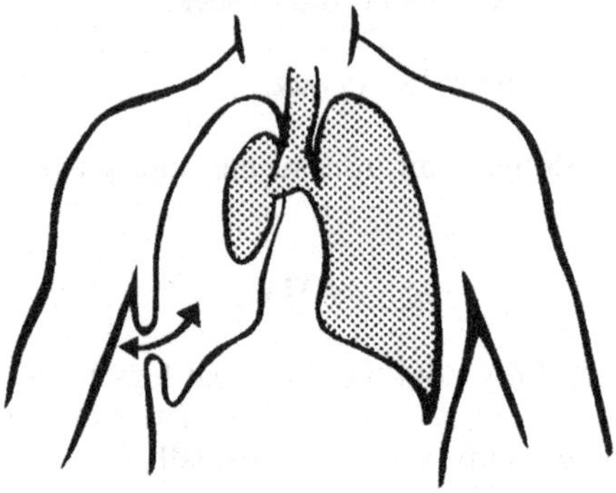

*Figure 3-1. Collapsed lung.*

## 3-5. First Aid for Chest Wounds

*a. Evaluate the Casualty.* Be prepared to perform first aid measures. These measures may include clearing the airway, rescue breathing, treatment for shock, and/or bleeding control.

*b. Expose the Wound.* If appropriate, cut or remove the casualty's clothing to expose the wound. Remember, **DO NOT** remove clothing that is stuck to the wound because additional injury may result. **DO NOT** attempt to clean the wound.

**NOTE**

Examine the casualty to see if there is an entry and exit wound. If there are two wounds (entry, exit), perform the same procedure for both wounds. Treat the more serious (heavier bleeding, larger) wound first. It may be necessary to improvise a dressing for the second wound by using strips of cloth, such as a torn T-shirt, or whatever material is available. Also, listen for sucking sounds to determine if the chest wall is punctured.

**CAUTION**

If there is an object impaled in the wound, **DO NOT** remove it. Apply a dressing around the object and use additional improvised bulky materials/dressings (use the cleanest materials available) to build up the area around the object. Apply a supporting bandage over the bulky materials to hold them in place.

**CAUTION**

**DO NOT REMOVE** protective clothing in a chemical environment. Apply dressings *over* the protective clothing.

*c. Open the Casualty's Field Dressing Plastic Wrapper.* In cases where there is a sucking chest wound, the plastic wrapper is used with the field dressing to create an occlusive dressing. If a plastic wrapper is not available, or if an additional wound needs to be treated; cellophane, foil, the casualty's poncho, or similar material may be used. The covering should be wide enough to extend 2 inches or more beyond the edges of the wound in all directions.

(1) Tear open one end of the casualty's plastic wrapper covering the field dressing. Be careful not to destroy the wrapper and **DO NOT** touch the inside of the wrapper.

(2) Remove the inner packet (field dressing).

(3) Complete tearing open the empty plastic wrapper using as much of the wrapper as possible to create a flat surface.

*d. Place the Wrapper Over the Wound.* Place the inside surface of the plastic wrapper directly over the wound *when the casualty exhales* and hold it in place (Figure 3-2). The casualty may hold the plastic wrapper in place if he is able.

*Figure 3-2. Open chest wound sealed with an occlusive dressing.*

*e.* *Apply the Dressing to the Wound.*

(1) Use your free hand and shake open the field dressing (Figure 3-3).

*Figure 3-3. Shaking open the field dressing.*

(2) Place the white side of the dressing on the plastic wrapper covering the wound (Figure 3-4).

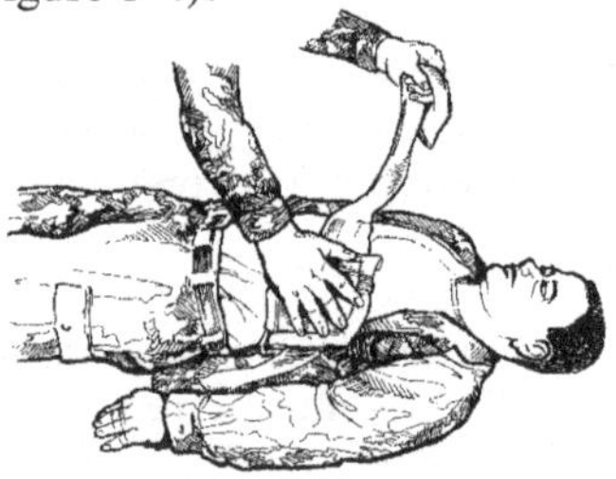

*Figure 3-4. Field dressing placed on plastic wrapper.*

**NOTE**

Use the casualty's field dressing, not your own.

(3) Have the casualty breathe normally.

(4) While maintaining pressure on the dressing, grasp one tail of the field dressing with the other hand and wrap it around the casualty's back. If tape is available, tape three sides of the plastic wrapper to the chest wall to provide occlusive type dressing. Leave one side untapped to provide emergency escape for air that may build up in the chest. If tape is not available, secure wrapper on three sides with field dressing leaving the fourth side as a flap.

(5) Wrap the other tail in the opposite direction, bringing both tails over the dressing (Figure 3-5).

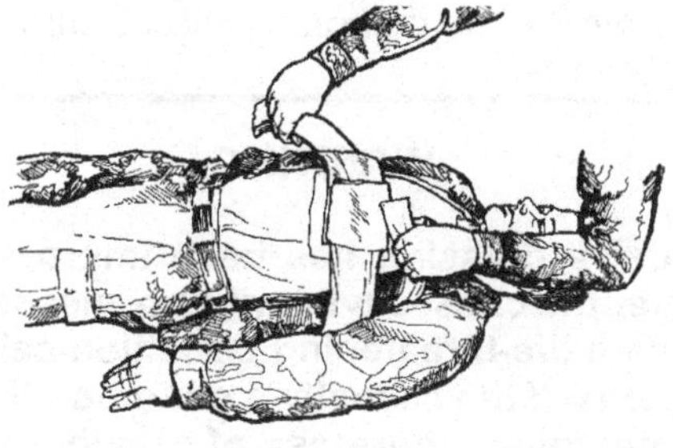

*Figure 3-5. Tails of field dressing wrapped around casualty in opposite direction.*

(6) Tie the tails into a square knot in the center of the dressing *after* the casualty exhales and *before* he inhales. This will aid in maintaining pressure on the bandage after it has been tied (Figure 3-6). Tie the dressing firmly enough to secure the dressing without interfering with the casualty's breathing.

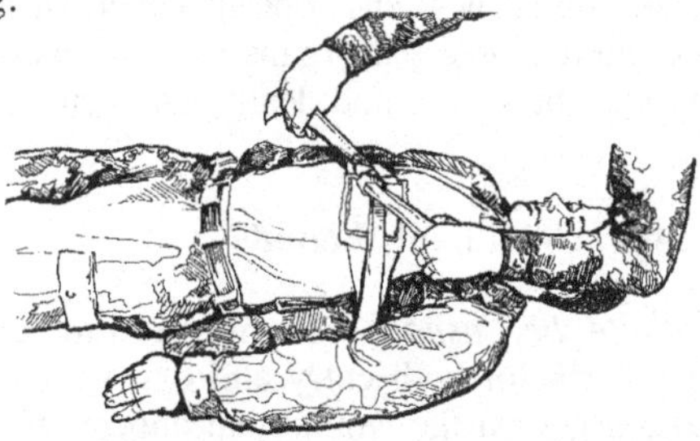

*Figure 3-6. Tails of dressing tied into square knot over center of dressing.*

**NOTE**

When practical, apply direct manual pressure over the dressing for 5 to 10 minutes to help control the bleeding.

*f.* *Position the Casualty*. Position the casualty on his injured side or in a sitting position, whichever makes breathing easier (Figure 3-7).

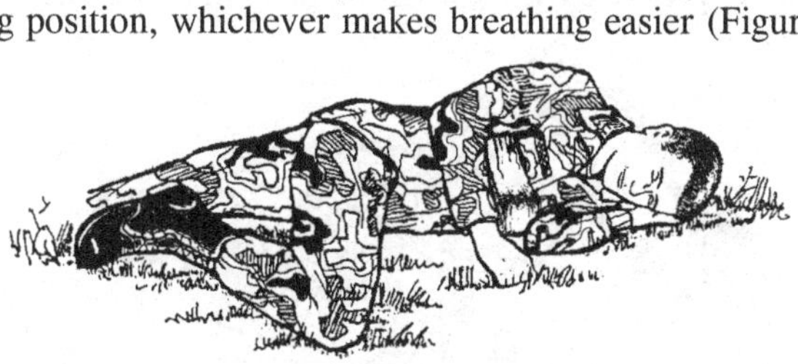

*Figure 3-7. Casualty positioned (lying) on injured side.*

*g. Seek Medical Assistance.* Contact medical personnel.

**WARNING**

**If an occlusive dressing has been improperly placed, air may enter the chest cavity with no means of escape. This causes a life-threatening condition called *tension pneumothorax*. If the casualty's condition (for example, difficulty breathing, shortness of breath, restlessness, or blueness/grayness of the skin) worsens after placing the dressing, quickly lift or remove, and then replace the occlusive dressing.**

## 3-6. Abdominal Wounds

The most serious abdominal wound is one in which an object penetrates the abdominal wall and pierces internal organs or large blood vessels. In these instances, bleeding may be severe and death can occur rapidly.

## 3-7. First Aid for Abdominal Wounds

*a. Evaluate the Casualty.* Be prepared to perform basic first aid measures. Always check for both entry and exit wounds. If there are two wounds (entry and exit), treat the wound that appears more serious first (for example, the heavier bleeding, protruding organs, larger wound, and so forth). It may be necessary to improvise dressings for the second wound by using strips of cloth, a T-shirt, or the cleanest material available.

*b. Position the Casualty.* Place and maintain the casualty on his back with his knees in an upright (flexed) position (Figure 3-8). The knees-up position helps relieve pain, assists in the treatment of shock, prevents further exposure of the bowel (intestines) or abdominal organs, and helps relieve abdominal pressure by allowing the abdominal muscles to relax.

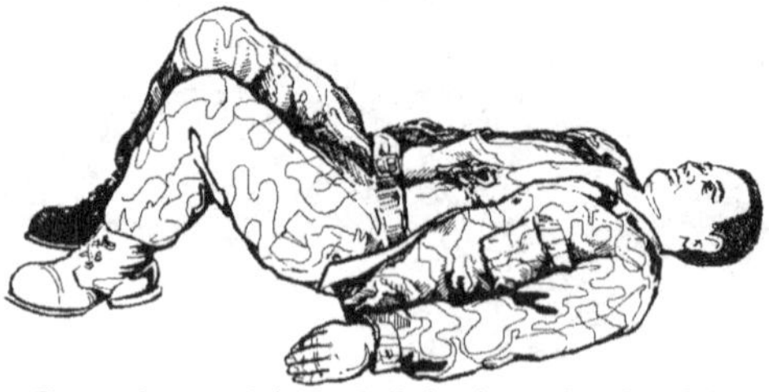

*Figure 3-8. Casualty positioned (lying) on back with knees (flexed) up.*

*c. Expose the Wound.*

(1) Remove the casualty's loose clothing to expose the wound. However, **DO NOT** attempt to remove clothing that is stuck to the wound; removing it may cause further injury.

**CAUTION**

**DO NOT REMOVE** protective clothing in a chemical environment. Apply dressings *over* the protective clothing.

(2) Gently pick up any organs that may be on the ground. Do this with a clean, dry dressing or with the cleanest available material. Place the organs on top of the casualty's abdomen (Figure 3-9).

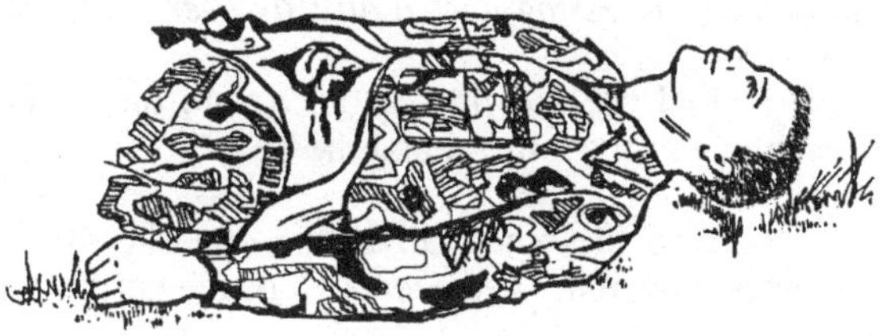

*Figure 3-9. Protruding organs placed near wound.*

NOTE

**DO NOT** probe, clean, or try to remove any foreign object from the abdomen. **DO NOT** touch with bare hands any exposed organs. **DO NOT** push organs back inside the body.

*d. Apply the Field Dressing.* Use the casualty's field dressing, not your own. If the field dressing is not large enough to cover the entire wound, the plastic wrapper from the dressing may be used to cover the wound first (placing the field dressing on top). Open the plastic wrapper carefully without touching the inner surface. If necessary, other improvised dressings may be made from clothing, blankets, or the cleanest materials available.

**WARNING**

**If there is an object extending from the wound, DO NOT remove it. Place as much of the wrapper over the wound as possible without dislodging or moving the object. DO NOT place the wrapper over the object.**

(1) Grasp the tails in both hands.

(2) Hold the dressing with the white side down directly over the wound. **DO NOT** touch the white (sterile) side of the dressing or allow anything except the wound to come in contact with it.

(3) Pull the dressing open and place it directly over the wound (Figure 3-10). If the casualty is able, he may hold the dressing in place.

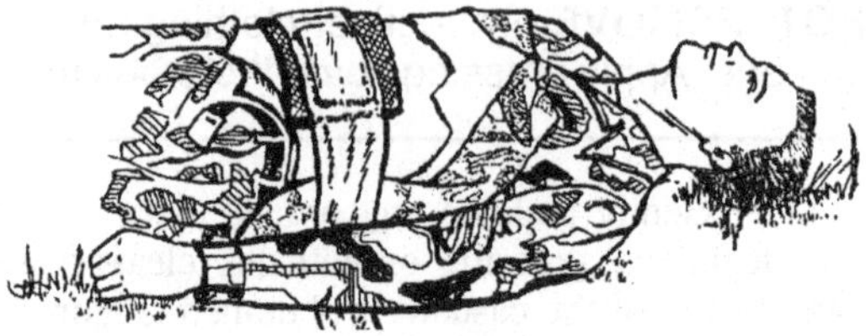

*Figure 3-10. Dressing placed directly over the wound.*

(4) Hold the dressing in place with one hand and use the other hand to wrap one of the tails around the body.

(5) Wrap the other tail in the opposite direction until the dressing is completely covered. Leave enough of the tail for a knot.

(6) Loosely tie the tails with a square knot at the casualty's side (Figure 3-11).

*Figure 3-11. Dressing applied and tails tied with a square knot.*

**WARNING**

**When the dressing is applied, DO NOT put pressure on the wound or exposed internal parts, because pressure could cause further injury (vomiting, ruptured intestines, and so forth). Therefore, tie the dressing ties (tails) loosely at casualty's side, not directly over the dressing.**

(7) Tie the dressing firmly enough to prevent slipping without applying pressure to the wound site (Figure 3-12).

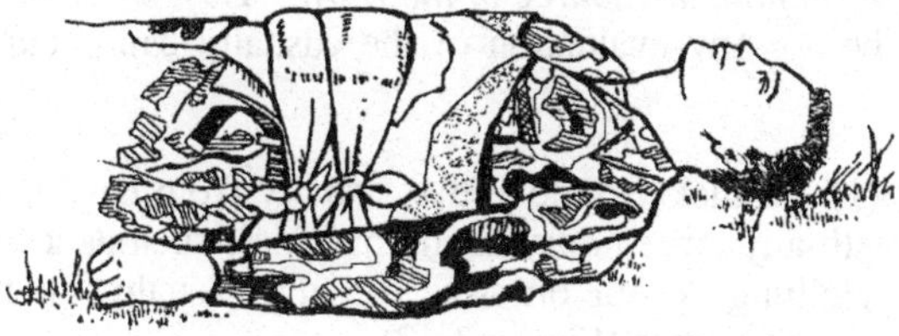

*Figure 3-12. Field dressing covered with improvised material and loosely tied.*

Field dressings can be covered with improvised reinforcement material (cravats, strips of torn T-shirt, or other cloth) for additional support and protection. Tie improvised bandage on the opposite side of the dressing ties firmly enough to prevent slipping but without applying additional pressure to the wound.

**CAUTION**

**DO NOT** give casualties with abdominal wounds food or water (moistening the lips is allowed).

*e. Seek Medical Assistance.* Notify medical personnel.

## 3-8. Burn Injuries

Burns often cause extreme pain, scarring, or even death. Before administering first aid, you must be able to recognize the type of burn. There are four types of burns:

- Thermal burns caused by fire, hot objects, hot liquids, and gases; or by nuclear blast or fireball.

- Electrical burns caused by electrical wires, current, or lightning.

- Chemical burns caused by contact with wet or dry chemicals or white phosphorus (WP)—from marking rounds and grenades.

- Laser burns (eye [ocular] injury).

## 3-9. First Aid for Burns

*a. Eliminate the Source of the Burn.* The source of the burn must be eliminated before any evaluation of the casualty can occur and first aid administered.

(1) Quickly remove the casualty from danger and cover the *thermal burn* with any large nonsynthetic material, such as a field jacket. If the casualty's clothing is still on fire, roll the casualty on the ground to smother (put out) the flames (Figure 3-13).

*Figure 3-13. Casualty covered and rolled on ground.*

> **CAUTION**
>
> Synthetic materials, such as nylon, may melt and cause further injury.

(2) Remove the *electrical burn* casualty from the electrical source by turning off the electrical current. **DO NOT** attempt to turn off the electricity if the source is not close by. Speed is critical, so **DO NOT** waste unnecessary time. If the electricity cannot be turned off, wrap any *nonconductive* material (*dry* rope, clothing, wood, and so forth) around the casualty's back and shoulders and drag the casualty away from the electrical source (Figure 3-14). **DO NOT** make body-to-body contact with the casualty or touch any wires because you could also become an electrical burn casualty.

*Figure 3-14. Casualty removed from electrical source (using nonconductive material).*

> **WARNING**
>
> **High voltage electrical burns may cause temporary unconsciousness, difficulties in breathing, or difficulties with the heart (heartbeat).**

(3) Remove the *chemical* from the *burned casualty*. Remove *liquid* chemicals by flushing with as much water as possible. Remove *dry* chemicals by brushing off loose particles (**DO NOT** use the bare surface of your hand because you could become a chemical burn casualty) and then flush with large amounts of water, if available. If large amounts of water are not available, then **NO** water should be applied because small amounts of water applied to a dry chemical burn may cause a chemical reaction. When WP strikes the skin, smother with a wet cloth or mud. Keep WP covered with a wet material to exclude air; this should help prevent the particles from burning.

(4) Remove the *laser burn* casualty from the source. When removing the casualty from the laser beam source, be careful not to enter the

beam or you may become a casualty. Never look directly at the beam source and if possible, wear appropriate eye protection.

NOTE

After the casualty is removed from the source of the burn, he should be evaluated for conditions requiring basic first aid measures.

*b. Expose the Burn.* Cut and gently lift away any clothing covering the burned area, without pulling clothing over the burns. Leave in place any clothing that is stuck to the burn. If the casualty's hands or wrists have been burned, remove jewelry if possible without causing further injury (rings, watches, and so forth) and place in his pockets. This prevents the necessity to cut off jewelry since swelling usually occurs as a result of a burn.

**CAUTION**

**DO NOT** lift or cut away clothing if in a chemical environment. Apply the dressing directly over the casualty's protective clothing. **DO NOT** attempt to decontaminate skin where blisters have formed.

*c. Apply a Field Dressing to the Burn.*

(1) Grasp the tails of the casualty's dressing in both hands.

(2) Hold the dressing directly over the wound with the white side down, pull the dressing open, and place it directly over the wound. **DO NOT** touch the white (sterile) side of the dressing or allow anything except the wound to come in contact with it. If the casualty is able, he may hold the dressing in place.

(3) Hold the dressing in place with one hand and use the other hand to wrap one of the tails around the limbs or the body.

(4) Wrap the other tail in the opposite direction until the dressing is completely covered.

(5) Tie the tails into a square knot over the outer edge of the dressing. The dressing should be applied lightly over the burn. Ensure that dressing is applied firmly enough to prevent it from slipping.

**NOTE**

Use the cleanest improvised dressing material available if a field dressing is not available or if it is not large enough for the entire wound.

*d.* *Take the Following Precautions*:

- **DO NOT** place the dressing over the face or genital area.
- **DO NOT** break the blisters.
- **DO NOT** apply grease or ointments to the burns.
- For electrical burns, check for both an entry and exit burn from the passage of electricity through the body. Exit burns may appear on any area of the body despite location of entry burn.
- For burns caused by wet or dry chemicals, flush the burns with large amounts of water and cover with a dry dressing.
- For burns caused by WP, flush the area with water, then cover with a wet material, dressing, or mud to exclude the air and keep the WP particles from burning.
- For laser burns, apply a field dressing.
- If the casualty is conscious and not nauseated, give him small amounts of water.

*e.* *Seek Medical Assistance*. Notify medical personnel.

## 3-10. Dressings and Bandages

*a.* *Head Wounds*.

(1) *Position the casualty*.

**WARNING**

**DO NOT move the casualty if you suspect he has sustained a neck, spine, or head injury (which produces any signs or symptoms other than minor bleeding).**

- If the casualty has a minor (superficial) scalp wound and is conscious:
  - Have the casualty sit up (unless other injuries prohibit or he is unable to).
  - If the casualty is lying down and is not accumulating fluids or drainage in his throat, elevate his head slightly.
  - If the casualty is bleeding from or into his mouth or throat, turn his head to the side or position him on his side so that the airway will be clear. Avoid putting pressure on the wound and place him on his uninjured side (Figure 3-15).

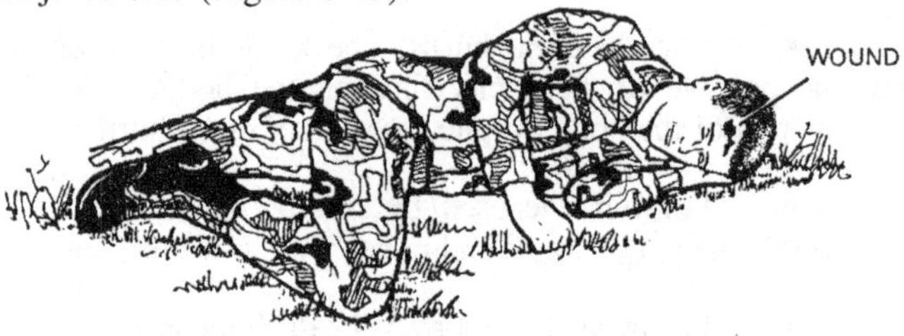


*Figure 3-15. Casualty lying on side opposite injury.*

- If the casualty is unconscious or has a severe head injury, then suspect and treat him as having a potential neck or spinal injury, *immobilize and **DO NOT** move the casualty.*

**NOTE**

If the casualty is choking or vomiting or is bleeding from or into his mouth (thus compromising his airway), position him on his uninjured side to allow for drainage and to help keep his airway clear.

**WARNING**

**If it is necessary to turn a casualty with a suspected neck/spine injury; roll the casualty gently onto his side, keeping the head, neck, and body aligned while providing support for the head and neck. DO NOT roll the casualty by yourself but seek assistance. *Move him only if absolutely necessary*, otherwise keep the casualty immobilized to prevent further damage to the neck/spine.**

(2) *Expose the wound.* Remove the casualty's helmet (if necessary). In a nuclear, biological, and chemical (NBC) environment, the first aid provider must leave the casualty as much protection (such as protective mask, mission-oriented protective posture [MOPP] overgarments) as possible. What items of protective equipment can be removed is dependent upon the casualty's injuries (where on the body and what type), the MOPP level, integrity of protective equipment (such as tears in the garment or mask seal), availability of chemical protective shelters, and the tactical situation.

**WARNING**

**DO NOT attempt to clean the wound or remove a protruding object.**

**NOTE**

Always use the casualty's field dressing, not your own.

(3) *Apply a dressing to a wound of the forehead or back of head.* To apply a dressing to a wound of the forehead or back of the head—

(*a*) Remove the dressing from the wrapper.

(*b*) Grasp the tails of the dressing in both hands.

(*c*) Hold the dressing (white side down) directly over the wound. **DO NOT** touch the white (sterile) side of the dressing or allow anything except the wound to come in contact with it.

(*d*) Place it directly over the wound.

(*e*) Hold it in place with one hand. If the casualty is able, he may assist.

(*f*) Wrap the first tail horizontally around the head; ensure the tail covers the dressing (Figure 3-16).

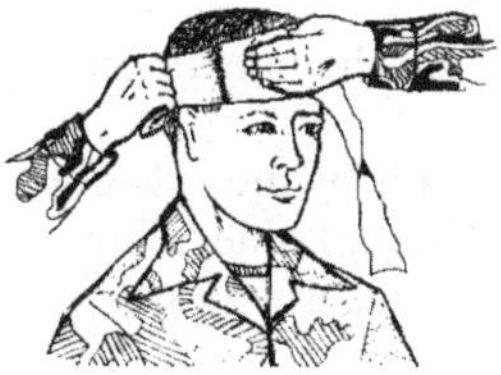

*Figure 3-16. First tail of dressing wrapped horizontally around head.*

(*g*) Hold the first tail in place and wrap the second tail in the opposite direction, covering the dressing (Figure 3-17).

*Figure 3-17. Second tail wrapped in opposite direction.*

(*h*) Tie a square knot and secure the tails at the side of the head, making sure they **DO NOT** cover the eyes or ears (Figure 3-18).

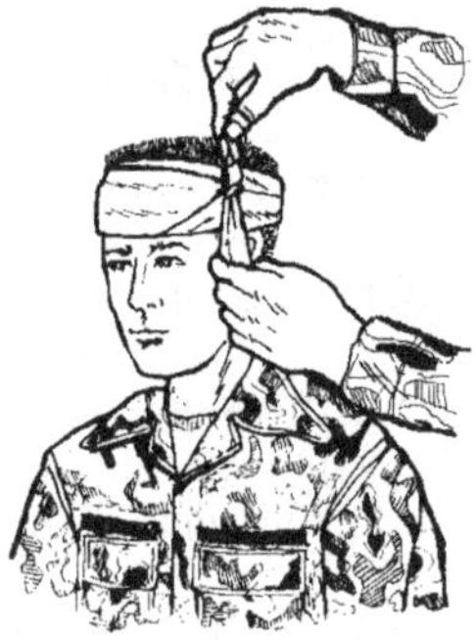

*Figure 3-18. Tails tied in square knot at side of head.*

(4) *Apply a dressing to a wound on top of the head.* To apply a dressing to a wound on top of the head—

(*a*) Remove the dressing from the wrapper.

(*b*) Grasp the tails of the dressing in both hands.

(*c*) Hold it (white side down) directly over the wound. **DO NOT** touch the white (sterile) side of the dressing or allow anything except the wound to come in contact with it.

(*d*) Place it over the wound (Figure 3-19).

*Figure 3-19. Dressing placed over wound.*

(*e*) Hold it in place with one hand. If the casualty is able, he may assist.

(*f*) Wrap one tail down under the chin (Figure 3-20), up in front of the ear, over the dressing, and in front of the other ear.

*Figure 3-20. One tail of dressing wrapped under chin.*

**WARNING**

**Ensure the tails remain wide and close to the front of the chin to avoid choking the casualty.**

(*g*) Wrap the remaining tail under the chin in the opposite direction and up the side of the face to meet the first tail (Figure 3-21).

*Figure 3-21. Remaining tail wrapped under chin in opposite direction.*

(*h*) Cross the tails (Figure 3-22), bringing one around the forehead (above the eyebrows) and the other around the back of the head (at the base of the skull) to a point just above and in front of the opposite ear, and tie them using a square knot (Figure 3-23).

*Figure 3-22. Tails of dressing crossed with one around forehead.*

*Figure 3-23. Tails tied in square knot (in front of and above ear).*

(5) *Apply a triangular bandage to the head.* To apply a triangular bandage to the head—

(*a*) Turn the base (longest side) of the bandage up and center its base on the center of the forehead, letting the point (apex) fall on the back of the neck (Figure 3-24A).

(*b*) Take the ends behind the head and cross the ends over the apex.

(*c*) Take them over the forehead and tie them (Figure 3-24B).

(*d*) Tuck the apex behind the crossed part of the bandage or secure it with a safety pin, if available (Figure 3-24C).

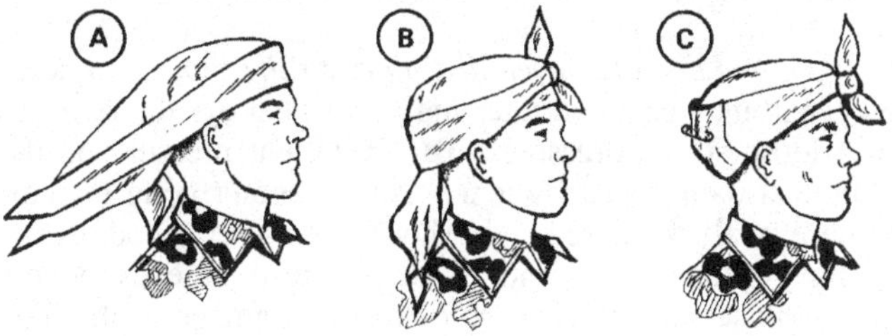


*Figure 3-24. Triangular bandage applied to head (Illustrated A—C)*

(6) *Apply a cravat bandage to the head.* To apply a cravat bandage to the head—

(*a*) Place the middle of the bandage over the dressing (Figure 3-25A).

(*b*) Cross the two ends of the bandage in opposite directions completely around the head (Figure 3-25B).

(*c*) Tie the ends over the dressing (Figure 3-25C).

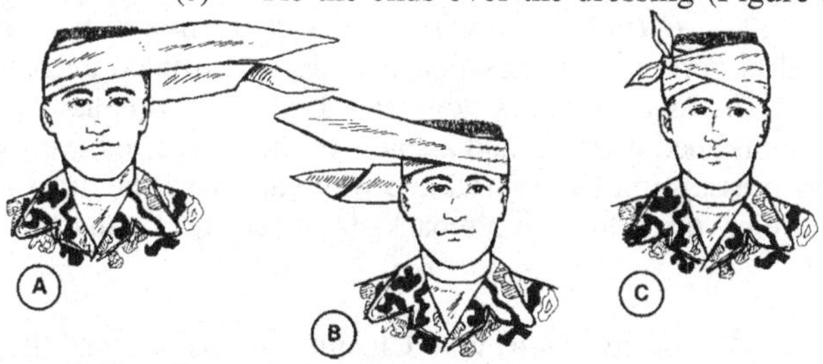


*Figure 3-25. Cravat bandage applied to head (Illustrated A—C).*

*b. Eye Injuries*. The eye is a vital sensory organ, and blindness is a severe physical handicap. Timely first aid of the eye may relieve pain and may also help to prevent shock, permanent eye injury, and possible loss of vision. Because the eye is very sensitive, any injury can be easily aggravated if it is improperly handled. Injuries of the eye may be quite severe. Cuts of the eyelids can appear to be very serious, but if the eyeball is not involved, a person's vision usually will not be damaged. However, lacerations (cuts) of the eyeball can cause permanent damage or loss of sight.

(1) *Lacerated/torn eyelids*. Lacerated eyelids may bleed heavily, but bleeding usually stops quickly. Cover the injured eye with a sterile dressing. **DO NOT** put pressure on the wound because you may injure the eyeball. Handle torn eyelids very carefully to prevent further injury. Place any detached pieces of the eyelid on a clean bandage or dressing and immediately send them with the casualty to the medical facility.

(2) *Lacerated eyeball (injury to the globe)*. Lacerations or cuts to the eyeball may cause serious and permanent eye damage. Cover the injury with a loose sterile dressing. **DO NOT** put pressure on the eyeball because additional damage may occur. An important point to remember is that when one eyeball is injured, you should immobilize both eyes. This is done by applying a bandage to both eyes. Because the eyes move together, covering both will lessen the chances of further damage to the injured eye. (However, in hazardous surroundings, leave uninjured eye uncovered to enable casualty to see.)

**CAUTION**

**DO NOT** apply pressure when there is a possible laceration of the eyeball. The eyeball contains fluid. Pressure applied over the eye will force the fluid out, resulting in permanent injury. **APPLY PROTECTIVE DRESSING WITHOUT ADDED PRESSURE**.

(3) *Extruded eyeballs*. Service members may encounter casualties with severe eye injuries that include an extruded eyeball (eyeball out-of-socket). In such instances you should gently cover the extruded eye with a loose moistened dressing and also cover the unaffected eye. **DO NOT** bind or exert pressure on the injured eye while applying the dressing. Keep the casualty quiet, place him on his back, treat for shock, and evacuate him immediately.

(4) *Burns of the eyes*. Chemical burns, thermal (heat) burns, and light burns can affect the eyes.

(*a*) *Chemical burns*. Injuries from chemical burns require immediate first aid. Mainly acids or alkalies cause chemical burns. The first aid measures consist of flushing the eyes immediately with large amounts of water for at least 5 to 20 minutes, or as long as necessary to flush out the chemical and, once flushed, bandaging the eyes. If the burn is an acid burn, you should flush the eye for at least 5 to 10 minutes. If the burn is an alkali burn, you should flush the eye for at least 20 minutes. After the eye has been flushed evacuate the casualty immediately.

(*b*) *Thermal burns*. When an individual suffers burns of the face from a fire, the eyes will close quickly due to extreme heat. This reaction is a natural reflex to protect the eyeballs; however, the eyelids remain exposed and are frequently burned. If a casualty receives burns of the eyelids or face—

- **DO NOT** apply a dressing.
- **DO NOT** touch.
- **SEEK** medical assistance immediately.

(*c*) *Light burns*. Exposure to intense light can burn an individual. Infrared rays, eclipse light (if the casualty has looked directly at the sun), or laser burns cause injuries of the exposed eyeball. Ultraviolet rays from arc welding can cause a superficial burn to the surface of the eye. These injuries are generally not painful but may cause permanent damage to the eyes. Immediate first aid is usually not required. Loosely bandaging the eyes may make the casualty more comfortable and protect his eyes from further injury caused by exposure to other bright lights or sunlight.

**CAUTION**

With impaled objects or significant sized foreign bodies, both eyes are usually bandaged to help secure the foreign body in the injured eye. In a battlefield environment, leave the uninjured eye uncovered so that the casualty can see.

*c*. *Side-of-Head or Cheek Wound*. Facial injuries to the side of the head or the cheek may bleed profusely (Figure 3-26). Prompt action is necessary to ensure that the airway remains open and also to control the bleeding. It may be necessary to apply a dressing. To apply a dressing—

(1) Remove the dressing from its wrapper.

(2) Grasp the tails in both hands.

(3) Hold the dressing directly over the wound with the white side down and place it directly on the wound (Figure 3-27A). **DO NOT** touch the white (sterile) side of the dressing or allow anything except the wound to come in contact with it.

(4) Hold the dressing in place with one hand (the casualty may assist if able). Wrap the top tail over the top of the head and bring it down in front of the ear (on the side opposite the wound), under the chin (Figure 3-27B) and up over the dressing to a point just above the ear (on the wound side).

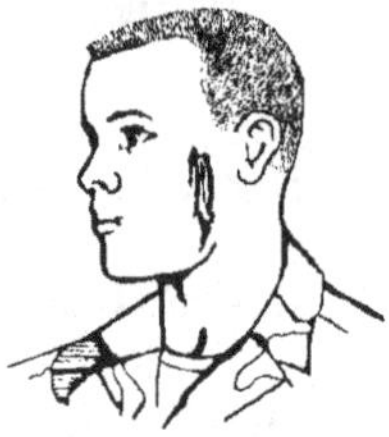

*Figure 3-26. Side of head or cheek wound.*

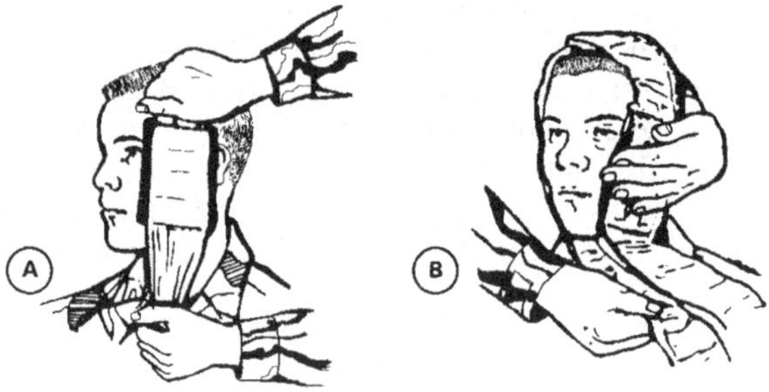


*Figure 3-27. Dressing placed directly on wound. Top tail wrapped over top of head, down in front of ear, and under chin (Illustrated A—B).*

NOTE

When possible, avoid covering the casualty's ear with the dressing, as this will decrease his ability to hear.

(5) Bring the second tail under the chin, up in front of the ear (on the side opposite the wound), and over the head to meet the other tail (on the wounded side) (Figure 3-28).

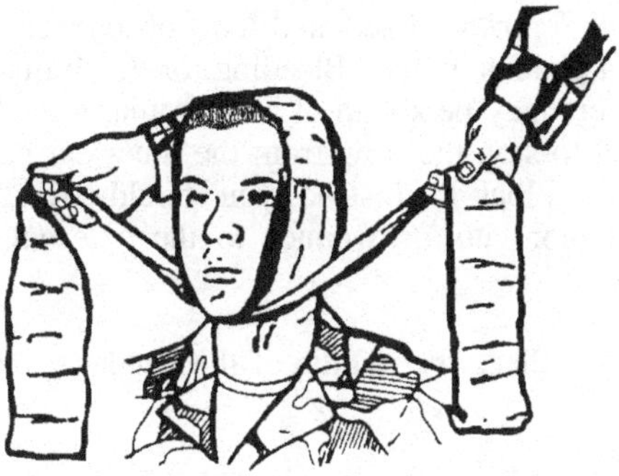

*Figure 3-28. Bringing second tail under the chin.*

(6) Cross the two tails (on the wound side) (Figure 3-29) and bring one end across the forehead (above the eyebrows) to a point just in front of the opposite ear (on the uninjured side).

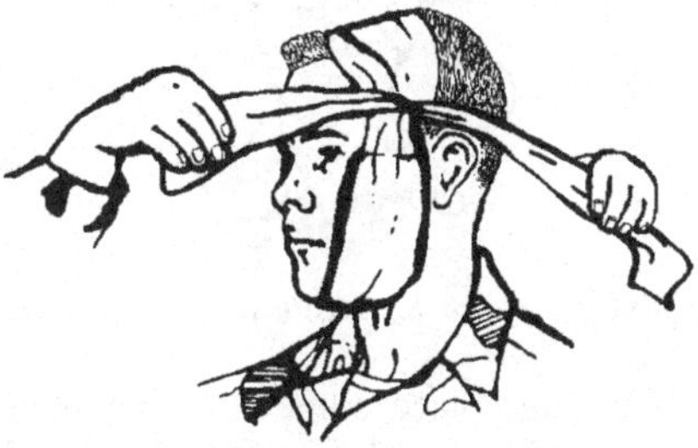

*Figure 3-29. Crossing the tails on the side of the wound.*

(7) Wrap the other tail around the back of the head (at the base of the skull), and tie the two ends just in front of the ear on the uninjured side with a square knot (Figure 3-30).

*Figure 3-30. Tying the tails of the dressing in a square knot.*

*d. Ear Injuries.* Lacerated (cut) or avulsed (torn) ear tissue may not, in itself, be a serious injury. Bleeding, or the drainage of fluids from the ear canal, however, may be a sign of a head injury, such as a skull fracture. **DO NOT** attempt to stop the flow from the inner ear canal nor put anything into the ear canal to block it. Instead, you should cover the ear lightly with a dressing. For minor cuts or wounds to the external ear, apply a cravat bandage as follows:

(1) Place the middle of the bandage over the ear (Figure 3-31A).

(2) Cross the ends, wrap them in opposite directions around the head, and tie them (Figures 3-31B and 3-31C).

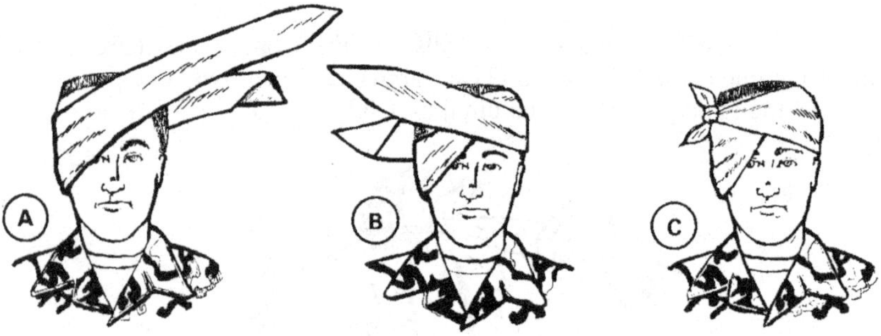


*Figure 3-31. Applying cravat bandage to ear (Illustrated A—C).*

(3) If possible, place some dressing material between the back of the ear and the side of the head to avoid crushing the ear against the head with the bandage.

*e. Nose Injuries.* Nose injuries generally produce bleeding. The bleeding may be controlled by placing an ice pack (if available) over the nose, or pinching the nostrils together. The bleeding may also be controlled by placing torn gauze (rolled) between the upper teeth and the lip.

> **CAUTION**
>
> **DO NOT** attempt to remove objects inhaled into the nose. An untrained person who removes such an object could worsen the casualty's condition and cause permanent injury.

*f. Jaw Injuries.* Before applying a bandage to a casualty's jaw, remove all loose or free-floating foreign material from the casualty's mouth.

If the casualty is unconscious, check for obstructions in the airway and remove if possible. If there is profuse bleeding in the oral cavity, the cavity may require loose packing with soft bandaging material (for example: Kerlix™ gauze) prior to applying a bandage. Care should be taken to avoid occluding the airway. When applying the bandage, allow the jaw enough freedom to permit passage of air and drainage from the mouth.

(1) *Apply bandages attached to field first aid dressing to the jaw.* After dressing the wound, apply the bandages using the same technique illustrated in Figure 3-32A—C.

**NOTE**

The dressing and bandaging procedure outlined for the jaw serves a twofold purpose. In addition to stopping the bleeding and protecting the wound, it also immobilizes a fractured jaw.

(2) *Apply a cravat bandage to the jaw.*

(*a*) Place the bandage under the chin and pull its ends upward. Adjust the bandage to make one end longer than the other (Figure 3-32A).

(*b*) Take the longer end over the top of the head to meet the short end at the temple and cross the ends over (Figure 3-32B).

(*c*) Take the ends in opposite directions to the other side of the head and tie them over the part of the bandage that was applied first (Figure 3-32C).

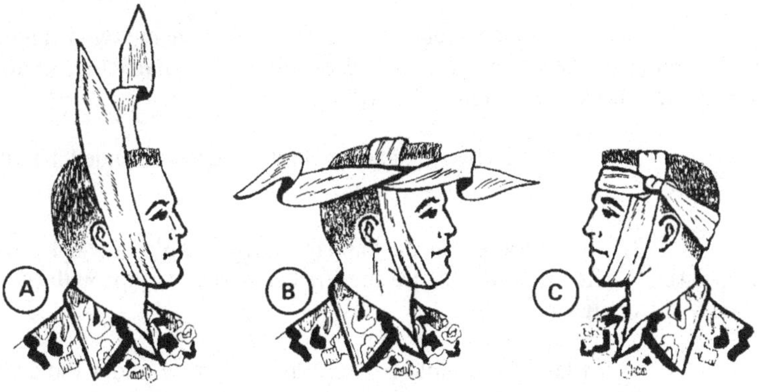


*Figure 3-32. Applying a cravat bandage to jaw (Illustrated A—C).*

NOTE

The cravat bandage technique is used to immobilize a fractured jaw or to maintain a sterile dressing that does not have tail bandages attached.

## 3-11. Shoulder Bandage

*a.* To apply bandages attached to the field first aid dressing—

(1) Take one bandage across the chest and the other across the back and under the arm opposite the injured shoulder.

(2) Tie the ends with a square knot (Figure 3-33).

*Figure 3-33. Shoulder bandage.*

*b.* To apply a cravat bandage to the shoulder or armpit—

(1) Make an extended cravat bandage by using two triangular bandages (Figure 3-34A); place the end of the first triangular bandage along the base of the second one (Figure 3-34B).

(2) Fold the two bandages into a single extended bandage (Figure 3-34C).

(3) Fold the extended bandage into a single cravat bandage (Figure 3-34D). After folding, secure the thicker part (overlap) with two or more safety pins (Figure 3-34E).

(4) Place the middle of the cravat bandage under the armpit so that the front end is longer than the back end and safety pins are on the outside (Figure 3-34F).

(5) Cross the ends on top of the shoulder (Figure 3-34G).

(6) Take one of the bandage ends across the back and under the arm on the opposite side and the other end across the chest. Tie the ends (Figure 3-34H).

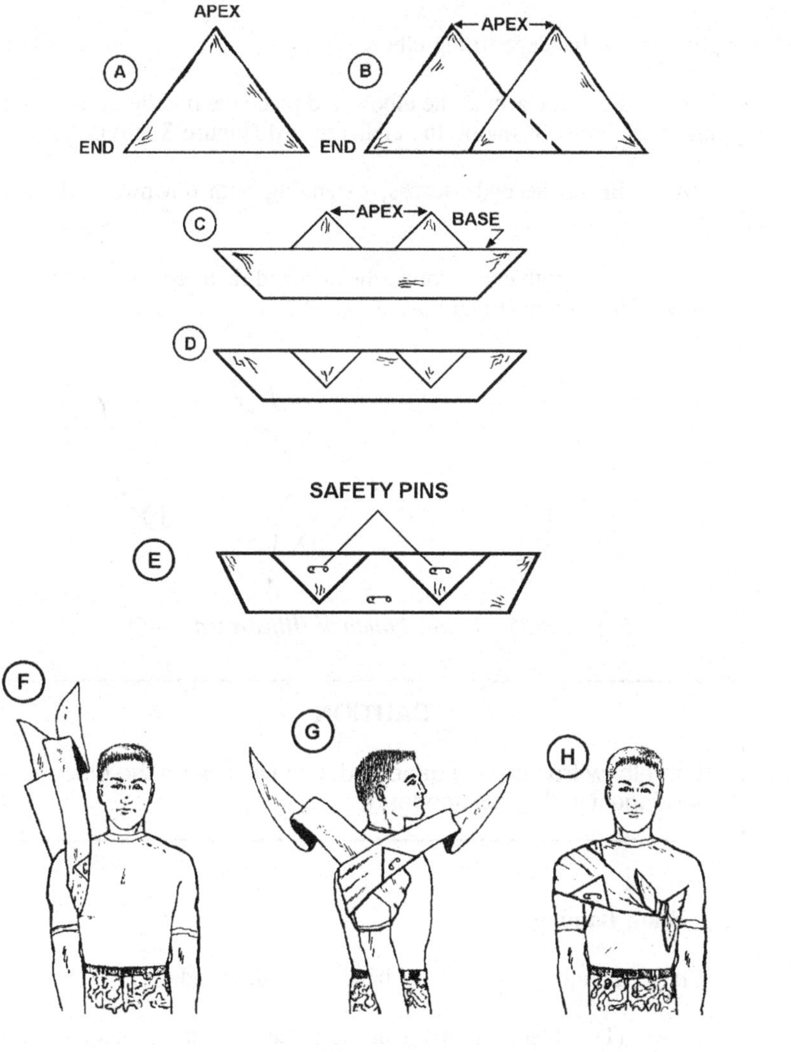


*Figure 3-34. Extended cravat bandage applied to shoulder or armpit (Illustrated A—H).*

Be sure to place sufficient wadding in the armpit. **DO NOT** tie the cravat bandage too tightly. Avoid compressing the major blood vessels in the armpit.

### 3-12. Elbow Bandage

To apply a cravat bandage to the elbow—

*a.* Bend the arm at the elbow and place the middle of the cravat at the point of the elbow bringing the ends upward (Figure 3-35A).

*b.* Bring the ends across, extending both downward (Figure 3-35B).

*c.* Take both ends around the arm and tie them with a square knot at the front of the elbow (Figure 3-35C).

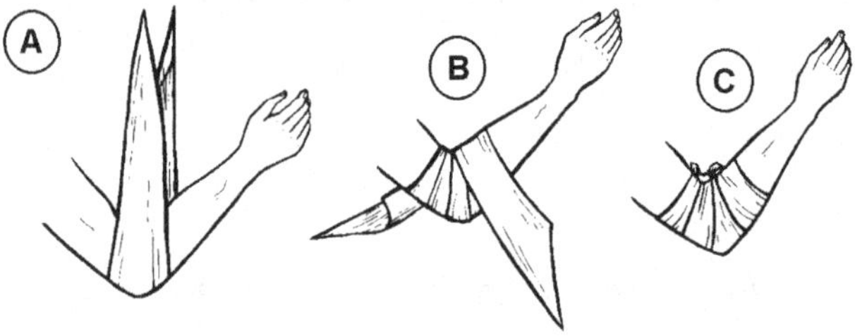


*Figure 3-35. Elbow bandage (Illustrated A—C).*

> **CAUTION**
>
> If an elbow fracture is suspected, **DO NOT** bend the elbow; bandage it in the position found.

### 3-13. Hand Bandage

*a.* To apply a triangular bandage to the hand—

(1) Place the hand in the middle of the triangular bandage with the wrist at the base of the bandage (Figure 3-36A). Ensure that the fingers are separated with absorbent material to prevent chafing and irritation of the skin.

(2) Place the apex over the fingers and tuck any excess material into the pleats on each side of the hand (Figure 3-36B).

(3) Cross the ends on top of the hand, take them around the wrist, and tie them (Figures 3-36C—E) with a square knot.

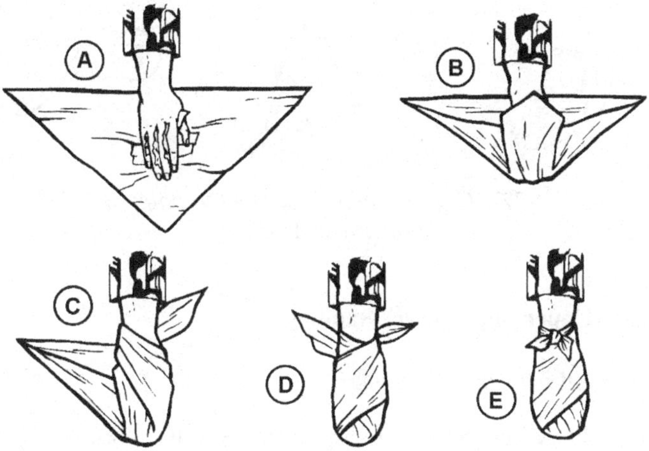


*Figure 3-36. Triangular bandage applied to hand (Illustrated A—E).*

*b*. To apply a cravat bandage to the palm of the hand—

(1) Lay the middle of the cravat over the palm of the hand with the ends hanging down on each side (Figure 3-37A).

(2) Take the end of the cravat at the little finger across the back of the hand, extending it upward over the base of the thumb; then bring it downward across the palm (Figure 3-37B).

(3) Take the thumb end across the back of the hand, over the palm, and through the hollow between the thumb and palm (Figure 3-37C).

(4) Take the ends to the back of the hand and cross them; then bring them up over the wrist and cross them again (Figure 3-37D).

(5) Bring both ends down and tie them with a square knot on top of the wrist (Figure 3-37E—F).

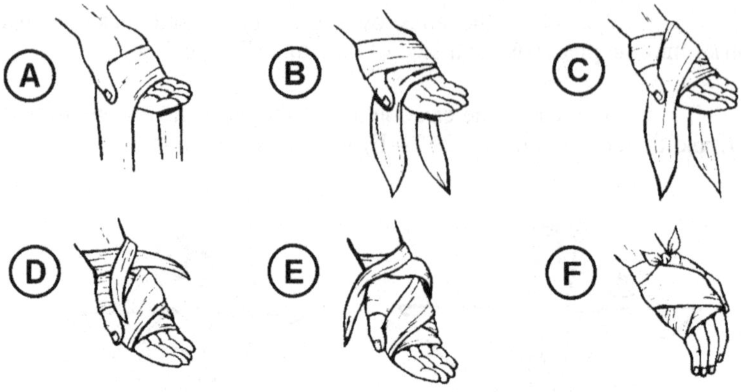


*Figure 3-37. Cravat bandage applied to palm of hand (Illustrated A—F).*

### 3-14. Leg (Upper and Lower) Bandage

To apply a cravat bandage to the leg—

*a.* Place the center of the cravat over the dressing (Figure 3-38A).

*b.* Take one end around and up the leg in a spiral motion and the other end around and down the leg in a spiral motion, overlapping part of each preceding turn (Figure 3-38B).

*c.* Bring both ends together and tie them (Figure 3-38C) with a square knot.

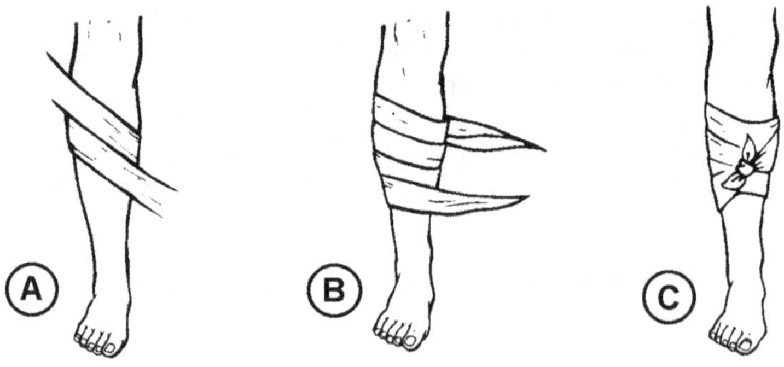


*Figure 3-38. Cravat bandage applied to leg (Illustrated A—C).*

## 3-15. Knee Bandage

To apply a cravat bandage to the knee as illustrated in Figure 3-39, use the same technique applied in bandaging the elbow.

**CAUTION**

If a fracture of the kneecap is suspected, **DO NOT** bend the knee; bandage it in the position found.

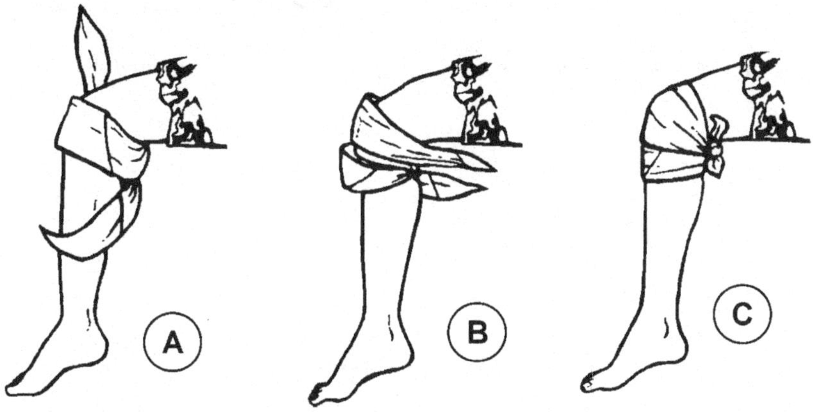


*Figure 3-39. Cravat bandage applied to knee (Illustrated A—C).*

## 3-16. Foot Bandage

To apply a triangular bandage to the foot—

*a.* Place the foot in the middle of the triangular bandage with the heel well forward of the base (Figure 3-40A). Ensure that the toes are separated by absorbent material to prevent chafing and irritation of the skin.

*b.* Place the apex over the top of the foot and tuck any excess material into the pleats on each side of the foot (Figure 3-40B).

*c.* Cross the ends on top of the foot, take them around the ankle, and tie them at the front of the ankle (Figure 3-40C—E).

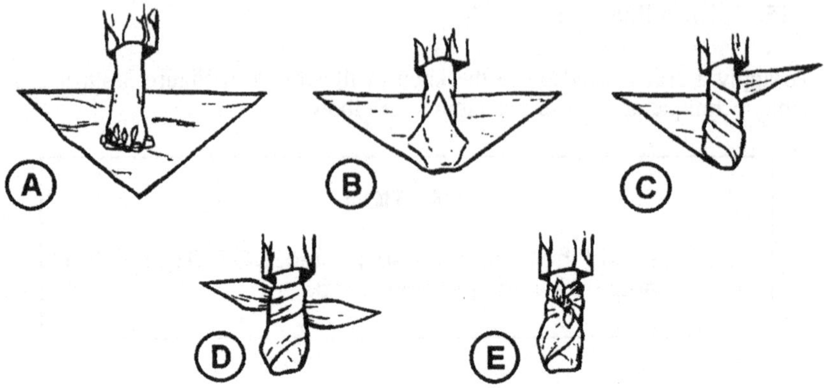


*Figure 3-40. Triangular bandage applied to foot (Illustrated A—E).*

CHAPTER 4

# FIRST AID FOR FRACTURES

## 4-1. General

A fracture is any break in the continuity of a bone. Fractures can cause total disability or in some cases death by severing vital organs and/or arteries. On the other hand, they can most often be treated so there is a complete recovery. The potential for recovery depends greatly upon the first aid the individual receives before he is moved. First aid includes immobilizing the fractured part in addition to applying lifesaving measures when necessary. The basic splinting principle is to immobilize the joints above and below the fracture.

## 4-2. Kinds of Fractures

Figure 4-1 depicts types of fractures.

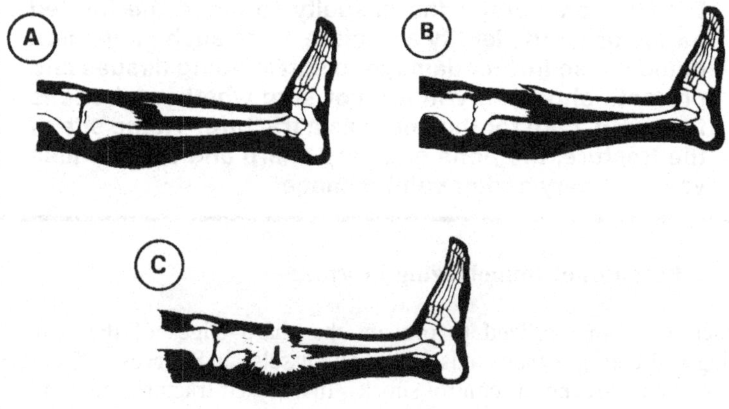


*Figure 4-1. Types of fractures (Illustrated A—C).*

*a.* *Closed Fracture* (Figure 4-1A). A closed fracture is a broken bone that does not break the overlying skin. The tissue beneath the skin may be damaged. A *dislocation* is when a joint, such as a knee, ankle, or shoulder, is not in the proper position. A *sprain* is when the connecting tissues of the joints have been torn. Dislocations and sprains (swelling, possible deformity, and discoloration) should be treated as closed fractures.

*b.* *Open Fracture* (Figure 4-1B and 4-1C). An open fracture is a broken bone that breaks (pierces) the overlying skin. The broken bone may come through the skin or a missile such as a bullet or shell fragment may go through the flesh and break the bone.

**NOTE**

An open fracture is contaminated and subject to infection.

## 4-3. Signs and Symptoms of Fractures

Indications of a fracture are deformity, tenderness, swelling, pain, inability to move the injured part, protruding bone, bleeding, or discolored skin at the injury site. A sharp pain when the service member attempts to move the part is also a sign of a fracture.

**WARNING**

**DO NOT encourage the casualty to move the injured part in order to identify a fracture since such movement could cause further damage to surrounding tissues and promote shock. If you are not sure whether a bone is fractured, care for the injury as a fracture. At the site of the fracture, the bone ends are sharp and could cause vessel (artery and/or vein) damage.**

## 4-4. Purposes of Immobilizing Fractures

A fracture is immobilized to prevent the sharp edges of the bone from moving and cutting tissue, muscle, blood vessels, and nerves. This reduces pain and helps prevent or control shock. In a closed fracture, immobilization keeps bone fragments from causing an open wound, which can become contaminated and subject to infection.

## 4-5. Splints, Padding, Bandages, Slings, and Swathes

*a. Splints.* Splints may be improvised from such items as boards, poles, sticks, tree limbs, or cardboard. If nothing is available for a splint, the chest wall can be used to immobilize a fractured arm and the uninjured leg can be used to immobilize (to some extent) the fractured leg.

*b. Padding.* Padding may be improvised from such items as a jacket, blanket, poncho, shelter half, or leafy vegetation.

*c.* *Bandages.* Bandages may be improvised from belts, rifle slings, kerchiefs, or strips torn from clothing or blankets. Narrow materials such as wire or cord should not be used to secure a splint in place. The application of wire and/or narrow material to an extremity could cause tissue damage and a tourniquet effect.

*d.* *Slings.* A sling is a bandage suspended from the neck to support an upper extremity. If a bandage is not available, a sling can be improvised by using the tail of a coat or shirt or pieces of cloth torn from such items as clothing and blankets. The triangular bandage is ideal for this purpose. Remember that the casualty's hand should be higher than his elbow, and the fingers should be showing at all times. The sling should be applied so that the supporting pressure is on the uninjured side.

*e.* *Swathes.* Swathes are any bands (pieces of cloth or load bearing equipment [LBE]) that are used to further immobilize a splinted fracture. Triangular and cravat bandages are often used and are called *swathe bandages*. The purpose of the swathe is to immobilize; therefore, the swathe bandage is placed above and/or below the fracture—not over it.

## 4-6. Procedures for Splinting Suspected Fractures

Before beginning first aid procedures for a fracture, gather whatever splinting materials are available. Ensure that splints are long enough to immobilize the joint above and below the suspected fracture. If possible, use at least four ties (two above and two below the fracture) to secure the splints. The ties should be square knots and should be tied away from the body on the splint. Distal pulses of the affected extremity should be checked before and after the application of the splint.

*a.* *Evaluate the Casualty.* Be prepared to perform any necessary lifesaving measures. Monitor the casualty for development of conditions that may require you to perform necessary lifesaving measures.

**WARNING**

**Unless there is immediate life-threatening danger, such as a fire or an explosion, DO NOT move the casualty with a suspected back or neck injury. Improper movement may cause permanent paralysis or death.**

**WARNING**

**In a chemical environment, DO NOT remove any protective clothing. Apply the dressings and splints over the garments.**

b. *Locate the Site of the Suspected Fracture.*

(1) Ask the casualty for the location of the injury.

- Does he have any pain?
- Where is it tender?
- Can he move the extremity?

NOTE

With the presence of an obvious deformity, do not make the casualty move extremity.

(2) Look for an unnatural position of the extremity.

(3) Look for a bone sticking out (protruding).

c. *Prepare the Casualty for Splinting the Suspected Fracture.*

(1) Reassure the casualty. Tell him that you will be providing first aid for him and that medical help is on the way.

(2) Loosen any tight or binding clothing.

(3) Remove all jewelry from the injured part and place it in the casualty's pocket. Tell the casualty you are doing this because if the jewelry is not removed and swelling occurs later, he may not be able to get it off and further bodily injury could result.

(4) Boots should not be removed from the casualty unless they are needed to stabilize a neck injury or there is actual bleeding from the foot.

d. *Gather Splinting Materials.* If standard splinting materials (splints, padding, and cravats) are not available, gather improvised materials. If splinting material is not available and the suspected fracture **CANNOT** be splinted, then swathes, or a combination of swathes and slings can be used to immobilize the extremity.

*e. Pad the Splints.* Pad the splints where they touch any bony part of the body, such as the elbow, wrist, knee, ankle, crotch, or armpit areas. Padding prevents excessive pressure on the area, which could lead to circulation problems.

*f. Check the Circulation Below the Site of the Injury.*

(1) Note any pale, white, or bluish-gray color of the skin, which may indicate impaired circulation. Circulation can also be checked by depressing the toe or fingernail beds and observing how quickly the color returns. A slower return of color to the injured side when compared with the uninjured side indicates a problem with circulation. The fingernail bed is the method to use to check the circulation in a dark-skinned casualty.

(2) Check the temperature of the injured extremity. Use your hand to compare the temperature of the injured side with the uninjured side. The body area below the injury may be colder to the touch indicating poor circulation.

(3) Question the casualty about the presence of numbness, tightness, cold, or tingling sensations.

**WARNING**

**Casualties with fractures of the extremities may show impaired circulation, such as numbness, tingling, cold or pale to bluish skin tone. These casualties should be evacuated by medical personnel and treated as soon as possible. Prompt medical treatment may prevent possible loss of the limb.**

**WARNING**

**If it is an open fracture and the bone is protruding from the skin, *DO NOT ATTEMPT TO PUSH THE BONE BACK UNDER THE SKIN.* Apply a field dressing over the wound to protect the area.**

*g. Apply the Splint in Place.*

(1) Splint the fracture in the position found. **DO NOT** attempt to reposition or straighten the injury. If it is an open fracture, stop the bleeding and protect the wound. Cover all wounds with field dressings before applying a splint. Remember to use the casualty's field dressing, not your own.

(2) Place one splint on each side of the fracture. Make sure that the splints reach, if possible, beyond the joints above and below the fracture.

(3) Tie the splints. Secure each splint in place above and below the fracture site with improvised (or actual) cravats. Improvised cravats, such as strips of cloth, belts, or whatever else you have, may be used. With minimal motion to the injured areas, place and tie the splints with the bandages. Push cravats through and under the natural body curvatures, and then gently position improvised cravats and tie in place. Use square knots. Tie all knots on the splint away from the casualty (Figure 4-2). **DO NOT** tie cravats directly over the suspected fracture site.

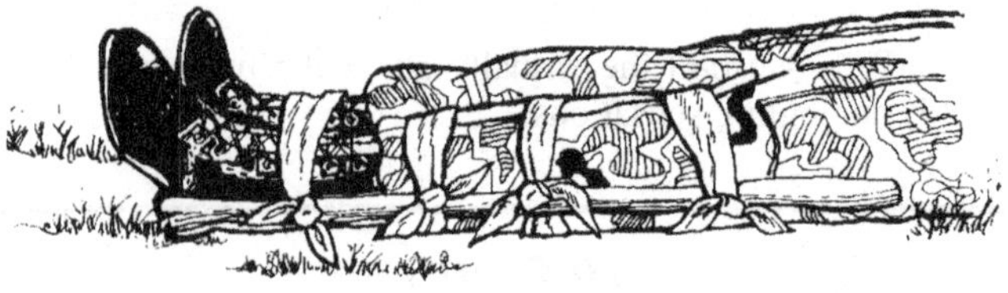

*Figure 4-2. Square knots tied away from casualty.*

*h. Check the Splint for Tightness.*

(1) **CHECK** to be sure that bandages are tight enough to securely hold splinting materials in place, but not so tight that circulation is impaired.

(2) **RECHECK** the circulation after application of the splint. Check the skin color and temperature. This is to ensure that the bandages holding the splint in place have not been tied too tightly. A fingertip check can be made by inserting the tip of the finger between the bandaged knot and the skin.

(3) **MAKE** any necessary adjustment without allowing the splint to become ineffective.

*i. Apply a Sling.* An improvised sling may be made from any available nonstretching piece of cloth, such as a battle dress uniform (BDU) shirt or trousers, poncho, or shelter half. Slings may also be improvised using the tail of a coat, belt, or a piece of cloth. Figure 4-3 depicts a shirttail used for support. A trousers belt or LBE may also be used for support (Figure 4-4). A sling should place the supporting pressure on the casualty's uninjured side. The supported arm should have the hand positioned slightly higher than the elbow showing the fingers.

*Figure 4-3. Shirttail used for support.*

*Figure 4-4. Belt used for support.*

(1) Insert the splinted arm in the center of the sling (Figure 4-5).

*Figure 4-5. Arm inserted in center of improvised sling.*

(2) Bring the ends of the sling up and tie them at the side (or hollow) of the neck on the uninjured side (Figure 4-6).

*Figure 4-6. Ends of improvised sling tied to side of neck.*

(3) Twist and tuck the corner of the sling at the elbow (Figure 4-7).

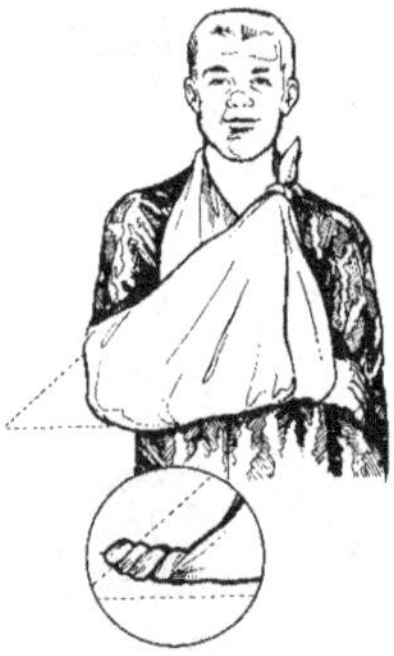

*Figure 4-7. Corner of sling twisted and tucked at elbow.*

*j. Apply a Swathe.* You may use any large piece of cloth, service member's belt, or pistol belt, to improvise a swathe.

**WARNING**

**The swathe should not be placed directly on top of the injury, but positioned either above or below the fracture site.**

(1) Apply swathes to the injured arm by wrapping the swathe over the injured arm, around the casualty's back, and under the arm on the uninjured side. Tie the ends on the uninjured side (Figure 4-8).

*Figure 4-8. Arm immobilized with strip of clothing.*

(2) A swathe is applied to an injured leg by wrapping the swathe around both legs and securing it on the uninjured side.

*k. Seek Medical Assistance.* Notify medical personnel, watch closely for development of life-threatening conditions and/or impaired circulation to the injured extremity. (Refer to Chapter 1 for additional information on life-threatening conditions.)

## 4-7. Upper Extremity Fractures

Figures 4-9 through 4-17 show how to apply slings, splints, and cravats (swathes) to immobilize and support fractures of the upper extremities. *Although the padding is not visible in some of the illustrations, it is always preferable to apply padding along the injured part for the length of the splint and especially where it touches any bony parts of the body.*

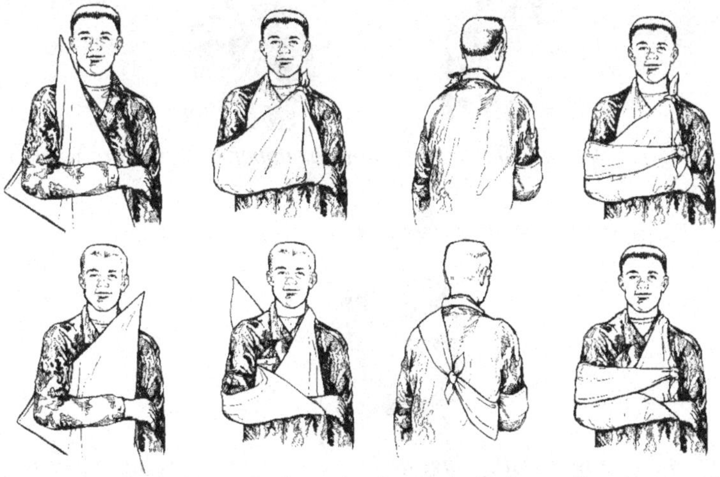

*Figure 4-9. Application of triangular bandage to form sling (two methods).*

*Figure 4-10. Completing sling sequence by twisting and tucking the corner of the sling at the elbow.*

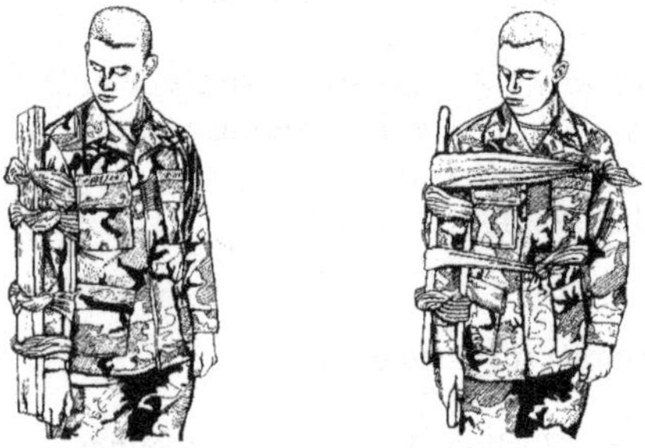

*Figure 4-11. Board splints applied to fractured elbow when elbow is not bent (two methods).*

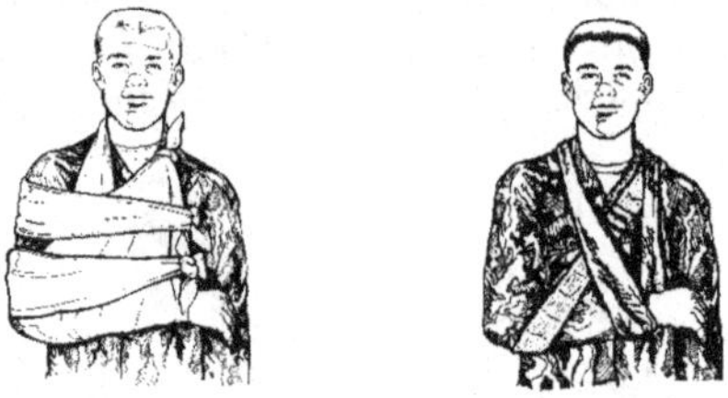

*Figure 4-12. Chest wall used as splint for upper arm fracture when no splint is available.*

*Figure 4-13. Chest wall, sling, and cravat used to immobilize fractured elbow when elbow is bent.*

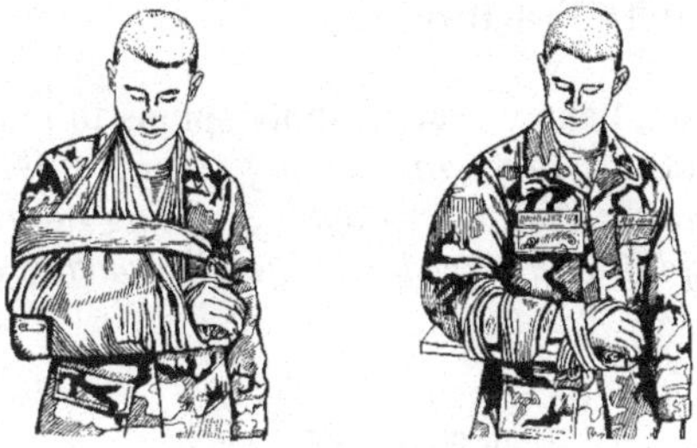

*Figure 4-14. Board splint applied to fractured forearm.*

*Figure 4-15. Fractured forearm or wrist splinted with sticks and supported with tail of shirt and strips of material.*

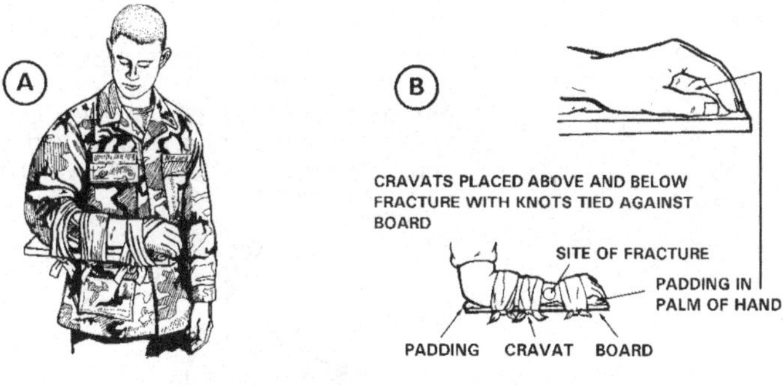


*Figure 4-16. Board splint applied to fractured wrist and hand (Illustrated A—B).*

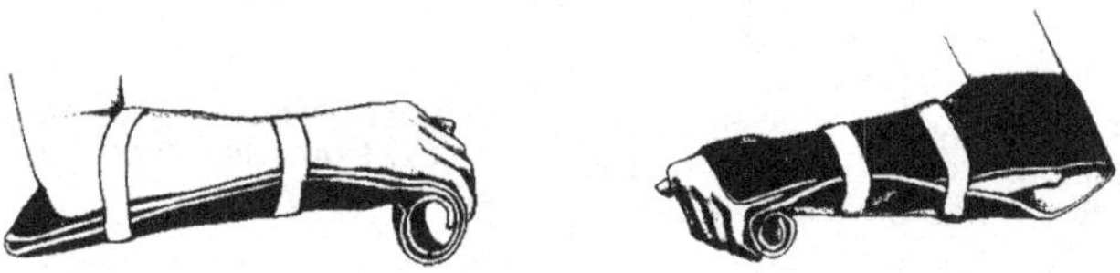

*Figure 4-17. SAM® splint applied to fractured wrist or forearm.*

## 4-8. Lower Extremity Fractures

Figures 4-18 through 4-24 show how to apply splints to immobilize fractures of the lower extremities. *Although padding is not visible in some of the illustrations, it is always preferable to apply padding along the injured part for the length of the splint and especially where it touches any bony parts of the body.*

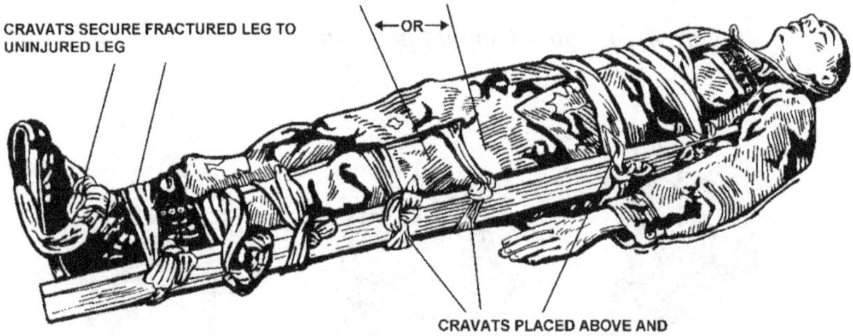


*Figure 4-18. Board splints applied to fractured hip or thigh.*

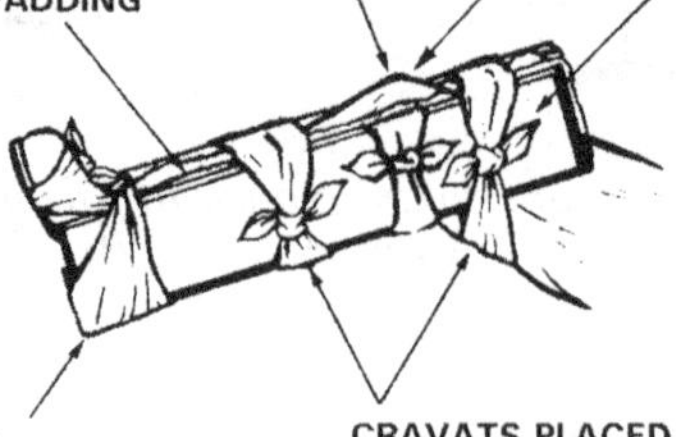


*Figure 4-19. Board splint applied to fractured or dislocated knee.*

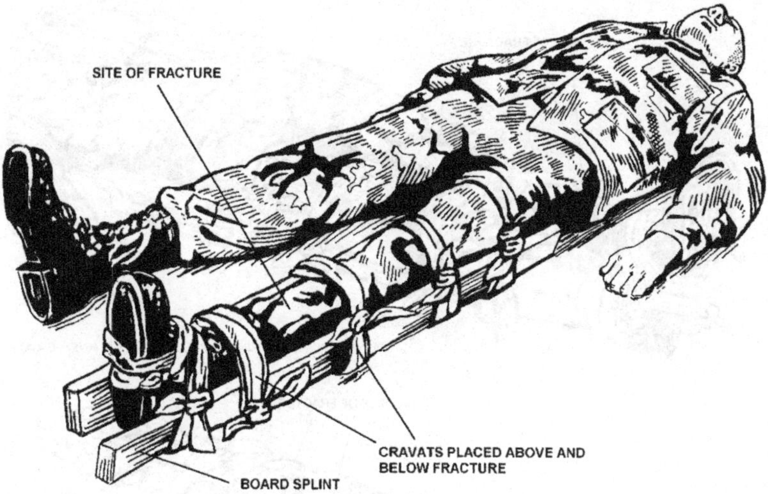


*Figure 4-20. Board splints applied to fractured lower leg or ankle.*

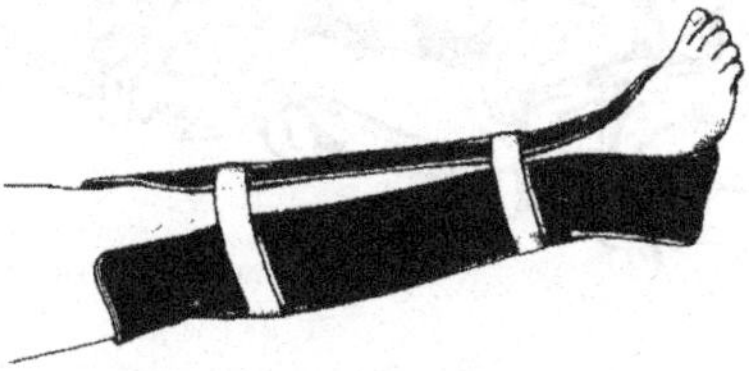

*Figure 4-21. SAM® splint applied to fractured lower leg or ankle.*

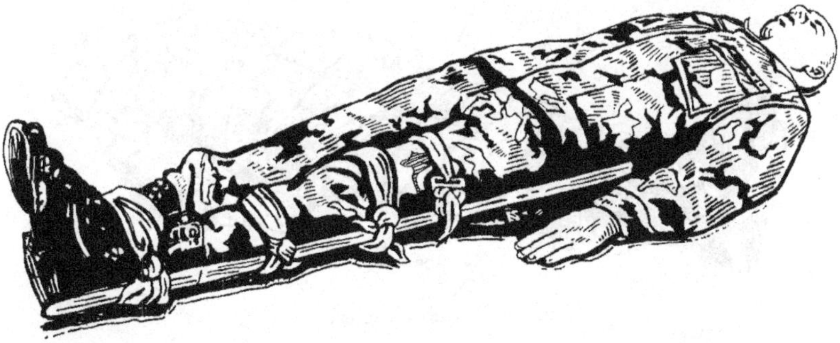

*Figure 4-22. Improvised splints applied to fractured lower leg or ankle.*

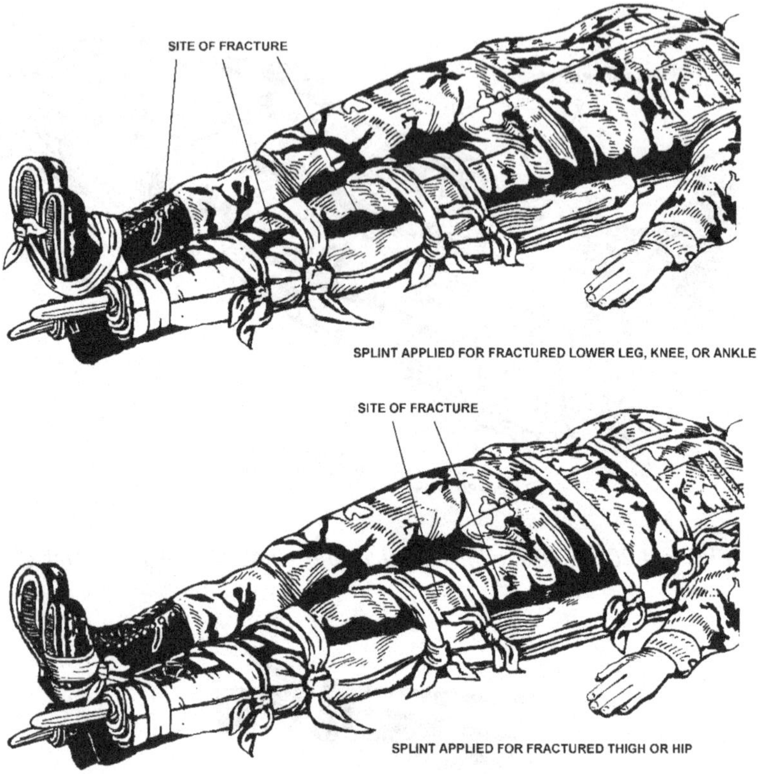


*Figure 4-23. Poles rolled in a blanket and used as splints applied to fractured lower extremity.*

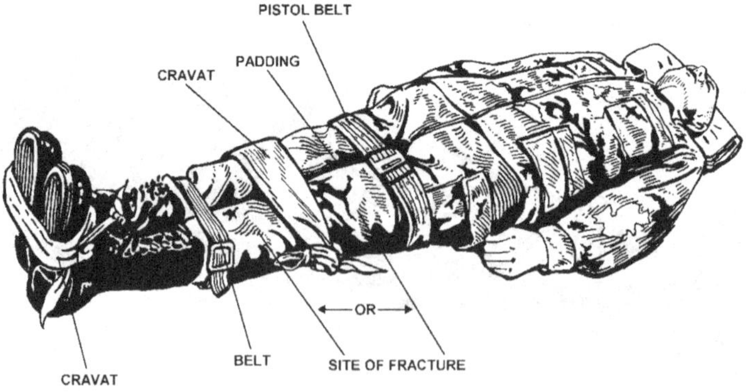


*Figure 4-24. Uninjured leg used as splint for fractured leg (anatomical splint).*

## 4-9. Jaw, Collarbone, and Shoulder Fractures

*a.* Apply a cravat to immobilize a fractured jaw as illustrated in Figure 4-25. Direct all bandaging support to the top of the casualty's head, not to the back of his neck. If incorrectly placed, the bandage will pull the casualty's jaw back and interfere with his breathing.

*Figure 4-25. Fractured jaw immobilized.*

**WARNING**

**Casualties with lower jaw (mandible) fractures cannot be laid flat on their backs because facial muscles will relax and may cause an airway obstruction.**

*b.* Apply two belts, a sling, and a cravat to immobilize a fractured collarbone, as illustrated in Figure 4-26.

*Figure 4-26. Application of belts, sling, and cravat to immobilize a fractured collarbone.*

*c.* Apply a sling and a cravat to immobilize a fractured or dislocated shoulder, using the technique illustrated in Figure 4-27.

*Figure 4-27. Application of sling and cravat to immobilize a fractured or dislocated shoulder.*

## 4-10. Spinal Column Fractures

It is often impossible to be sure a casualty has a fractured spinal column. Be suspicious of any back injury, especially if the casualty has fallen or if his back has been sharply struck or bent. If a casualty has received such an injury and does not have feeling in his legs or cannot move them, you can be reasonably sure that he has a severe back injury, which should be managed as a fracture. Remember, that the possibility of a neck fracture or injury to the back should always be suspected, and it is often impossible to be sure if a casualty has a fractured spinal column. If the spine is fractured, bending it can cause the sharp bone fragments to bruise or cut the spinal cord and result in permanent paralysis or death (Figure 4-28A). The spinal column must maintain normal spinal position at the lower back (lumbar region) to help remove pressure from the spinal cord.

*a.* If the casualty is not to be transported until medical personnel arrive—

- Caution him not to move. Ask him if he is in pain or if he is unable to move any part of his body.

- Leave him in the position in which he is found. **DO NOT** move any part of his body, unless he is in imminent danger.

- If the casualty is lying face up, slip a blanket or other supporting material under the arch of his lower back to help support the spine in a normal position (Figure 4-28B). Take care not to place so much bulky padding as to cause potential damage by causing undo pressure on the spine. If he is lying face down, **DO NOT** put anything under any part of his body.

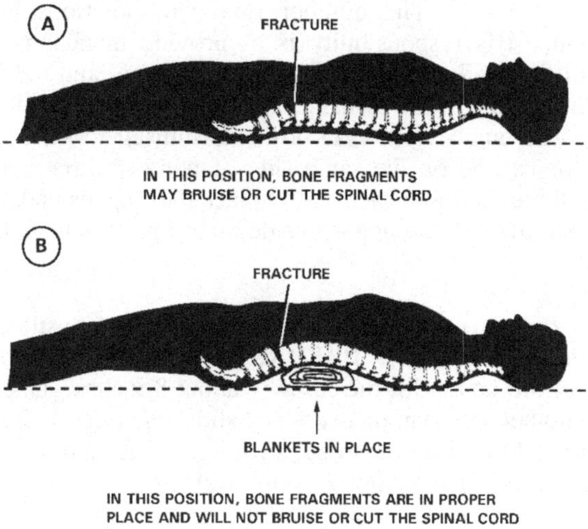


*Figure 4-28. Spinal column must maintain a normal spine position.*

*b.* If the casualty must be transported to a safe location before medical personnel arrive and if the casualty is in a—

- Face-up position, transport him by litter or use a firm substitute, such as a wide board or a door longer than his height. Loosely tie the casualty's wrists together over his waistline, using a cravat or a strip of cloth. Tie his feet together to prevent the accidental dropping or shifting of his legs. Lay a folded blanket across the litter where the arch of his back is to be placed. Using a four-man team (Figure 4-29), place the casualty on the litter without bending his spinal column or his neck.

*Figure 4-29. Placing face-up casualty with fractured back onto litter.*

- The number two man positions himself at the casualty's head. His responsibility is to provide manual in-line (neutral) stabilization of the head and neck. The number three, and four men position themselves on one side of the casualty; all kneel on one knee along the side of the casualty. The number one man positions himself to the opposite side of the casualty (or can be on the same side of number three and four). The numbers two, three, and four men gently place their hands under the casualty. The number one man on the opposite side places his hands under the injured part to assist.

- When all four men are in position to lift, the number two man commands, "**PREPARE TO LIFT**" and then, "**LIFT**." All men, in unison, gently lift the casualty about 8 inches. Once the casualty is lifted, the number one man recovers and slides the litter under the casualty, ensuring that the blanket is in proper position. The number one man then returns to his original lift position (Figure 4-29).

- When the number two man commands, "**LOWER CASUALTY**," all men, in unison, gently lower the casualty onto the litter.

- Facedown position, he must be transported in this same position. The four-man team lifts him onto a regular or improvised litter, keeping the spinal column in a normal spinal position. If a regular litter is used, first place a folded blanket on the litter at the point where the chest will be placed.

## 4-11. Neck Fractures

A fractured neck is extremely dangerous. Bone fragments may bruise or cut the spinal cord just as they might in a fractured back.

*a.* If the casualty is not to be transported until medical personnel arrive—

- Caution him not to move. Moving may cause permanent injury or death.

- Leave the casualty in the position in which he is found. If his neck and head (cervical spine) are in an abnormal position, immediately immobilize the neck and head.

- Keep his head still, if the casualty is lying face up, raise his shoulders slightly, and slip a roll of cloth that has the bulk of a bath towel under his neck (Figure 4-31). The roll should be thick enough to arch his neck only slightly, leaving the back of his head on the ground. **DO NOT** bend his neck or head forward. **DO NOT** raise or twist his head. Immobilize the casualty's head (Figure 4-32). Do this by padding heavy objects (such as rocks or his boots filled with dirt, sand, gravel, or rock) and placing them on each side of his head. If it is necessary to use boots, after filling them, tie the top tightly or stuff with pieces of cloth to secure the contents.)

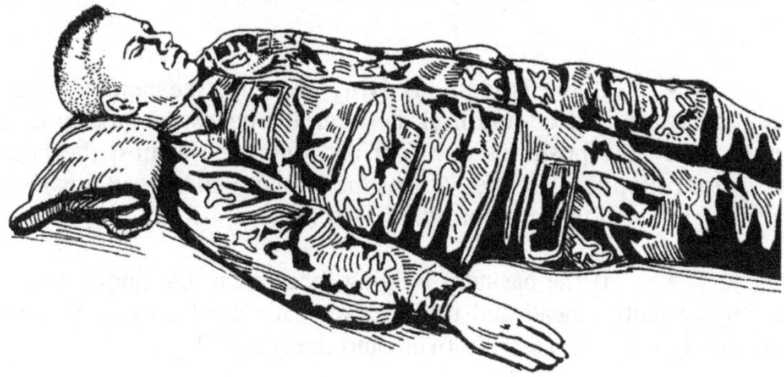

*Figure 4-30. Casualty with roll of cloth (bulk) under neck.*

*Figure 4-31. Immobilization of fractured neck.*

- **DO NOT** move him if the casualty is lying face down. Immobilize the head and neck by padding heavy objects and placing them on each side of his head. DO NOT put a roll of cloth under the neck. **DO NOT** bend the neck or head, nor roll the casualty onto his back.

*b.* If the casualty must be prepared for transportation before medical personnel arrive—

- If the casualty has a fractured neck, at least two persons are needed because the casualty's head and trunk must be moved in unison. The two persons must work in close coordination (Figure 4-32) to avoid bending of the neck.

- A wide board is placed lengthwise beside the casualty. It should extend at least 4 inches beyond the casualty's head and feet (Figure 4-32A).

- If the casualty is lying face up, the number one man steadies the casualty's head and neck between his hands. At the same time, the number two man positions one foot and one knee against the board to prevent it from slipping. He then grasps the casualty underneath his shoulder and hip and gently slides him onto the board (Figure 4-32B).

- If the casualty is lying face down, the number one man steadies the casualty's head and neck between his hands, while the number two man gently rolls the casualty over onto the board (Figure 4-32C).

- The number one man continues to steady the casualty's head and neck. The number two man simultaneously raises the casualty's shoulders slightly, places padding under his neck, and immobilizes the casualty's head (Figures 4-32D—E).

- Any improvised supports are secured in position with a cravat or strip of cloth extended across the casualty's forehead and under the board (Figure 4-32D).

- The board is lifted onto a litter or blanket in order to transport the casualty (Figure 4-32E).

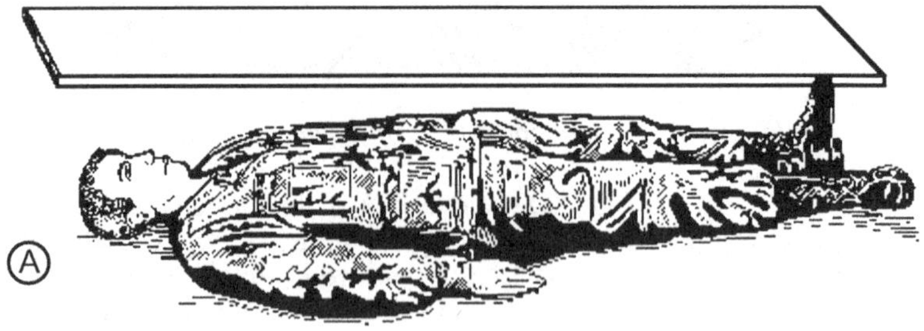


*Figure 4-32. Preparing casualty with fractured neck for transportation (Illustrated A—E).*

*Figure 4-32. Preparing casualty with fractured neck for transportation (Illustrated A—E) (Continued).*

CHAPTER 5

# FIRST AID FOR CLIMATIC INJURIES

## 5-1. General

*a.* It is desirable, but not always possible, for an individual's body to become adjusted (acclimated) to an environment.

(1) The service members physical condition determines the amount of time their bodies need to adjust to the environment. Even those individuals in good physical condition need time before working or training in extremes of hot or cold weather. Climate-related injuries are usually preventable; prevention is both an individual and leadership responsibility.

(2) Several factors contribute to health and well-being in any environment—

- Diet.
- Sleep and rest.
- Exercise.
- Suitable clothing.

(3) Diet should be suited to an individual's needs in a particular climate. A special diet started for any purpose (such as weight reduction) should be done with appropriate medical supervision.

**WARNING**

**Service members should use extreme caution when starting fad diets or taking over-the-counter herbal supplements. Medical records revealed that deaths and severe injuries occurred in individuals using dietary/ herbal supplements without medical monitoring.**

**NOTE**

Weight loss and the use of weight loss supplements should be supervised by a trained health care provider.

(4) Specialized clothing and equipment (such as cold weather gear) for a specific environment should be obtained and used properly.

*b.* For information on the prevention of heat and cold injuries, refer to FM 21-10/Marine Corps Reference Publication (MCRP) 4-11.1D.

## 5-2. Heat Injuries

*a.* Heat injuries are environmental injuries. They may result when a service member—

- Is exposed to extreme heat, such as from the sun or from high temperatures.
- Does not wear proper clothing.
- Is in MOPP gear.
- Is inside closed spaces, such as inside an armored vehicle.
- Wears body armor.

*b.* Heat injury can be divided into three categories: heat cramps, heat exhaustion, and heatstroke.

*c.* Each service member must be able to recognize and give first aid for heat injuries.

**WARNING**

**The heat casualty should be continually monitored for development of conditions which may require the performance of necessary basic lifesaving measures.**

**CAUTION**

Do not use salt solutions in first aid procedures for heat injuries.

(1) Check the casualty for signs and symptoms of *cramping*.

- *Signs and symptoms.* Cramping is caused by an imbalance of chemicals (called electrolytes) in the body as a result of excessive sweating. This condition causes the casualty to exhibit:

- Cramping in the extremities (arms and legs).
- Abdominal (stomach) cramps.
- Excessive sweating.

**NOTE**

Thirst may or may not occur. Cramping can occur without the service member being thirsty.

- *First aid measures.*

- Move the casualty to a cool, shady area or improvise shade if none is available.

- Loosen his clothing (if not in a chemical environment).

**NOTE**

In a chemical environment, transport the heat casualty to a noncontaminated area as soon as the mission permits.

- Have him slowly drink at least one canteen full of water. (The body absorbs cool water faster than warm or cold water; therefore, cool water is preferred if it is available.)

- Seek medical assistance should cramps continue.

(2) Check the casualty for signs and symptoms of *heat exhaustion*.

- *Signs and symptoms*. Heat exhaustion is caused by loss of body fluids (dehydration) through sweating without adequate fluid replacement. It can occur in an otherwise fit individual who is involved in physical exertion in any hot environment especially if the service member is not acclimatized to that environment. These signs and symptoms are—

- Excessive sweating with pale, moist, cool skin.
- Headache.
- Weakness.

- Dizziness.
- Loss of appetite.
- Cramping.
- Nausea (with or without vomiting).
- Urge to defecate.
- Chills (gooseflesh).
- Rapid breathing.
- Tingling of hands and/or feet.
- Confusion.

- *First aid measures.*

  - Move the casualty to a cool, shady area or improvise shade if none is available.
  - Loosen or remove his clothing and boots (unless in a chemical environment); pour water on him and fan him.
  - Have him slowly drink at least one canteen of water.
  - Elevate his legs.
  - If possible, the casualty should not participate in strenuous activity for the remainder of the day.
  - Monitor the casualty until the symptoms are gone, or medical assistance arrives.

(3) Check the casualty for signs and symptoms of *heatstroke*.

**WARNING**

**Heatstroke is a medical emergency which may result in death if care is delayed.**

- *Signs and symptoms.* A service member suffering from heatstroke has been exposed to high temperatures (such as direct sunlight) or been dressed in protective overgarments, which causes the body temperature to rise. Heatstroke occurs more rapidly in service members who are engaged in work or other physical activity in a high heat environment. Heatstroke is caused by a failure of the body's cooling mechanism which includes a decrease in the body's ability to produce sweat. The casualty's skin is red (flushed), hot, and dry. He may experience weakness, dizziness, confusion, headaches, seizures, nausea, stomach pains or cramps, and his respiration and pulse may be rapid and weak. Unconsciousness and collapse may occur suddenly.

- *First aid measures.* Cool casualty immediately by—

  - Moving him to a cool, shady area or improvising shade if none is available.

  - Loosening or removing his clothing (except in a chemical environment).

  - Spraying or pouring water on him; fanning him to permit the coolant effect of evaporation.

  - Massaging his extremities and skin, which increases the blood flow to those body areas, thus aiding the cooling process.

  - Elevating his legs.

  - Having him slowly drink at least one canteen full of water if he is conscious.

**NOTE**

Start cooling casualty immediately. Continue cooling while awaiting transportation and during transport to an MTF.

- *Medical assistance.* Seek medical assistance because the casualty should be transported to an MTF as soon as possible. Do not interrupt the cooling process or lifesaving measures to seek help; if someone else is present send them for help. The casualty should be continually monitored for development of conditions that may require the performance of necessary basic lifesaving measures.

*d.* *Table.* See Table 5-1 for further information.

*Table 5-1. Heat Injuries*

| INJURIES | SIGNS AND SYMPTOMS | FIRST AID[1] |
|---|---|---|
| **HEAT CRAMPS** | THE CASUALTY EXPERIENCES MUSCLE CRAMPS OF THE ARMS, LEGS, AND/OR STOMACH. THE CASUALTY MAY ALSO HAVE EXCESSIVE SWEATING. | 1. MOVE THE CASUALTY TO A COOL SHADY AREA OR IMPROVISE SHADE AND LOOSEN CLOTHING.[2]<br><br>2. HAVE HIM SLOWLY DRINK AT LEAST ONE CANTEEN FULL OF COOL WATER SLOWLY.<br><br>3. MONITOR THE CASUALTY AND GIVE HIM MORE WATER AS TOLERATED. |
| **HEAT EXHAUSTION** | THE CASUALTY EXPERIENCES HEAVY SWEATING WITH PALE, MOIST, COOL SKIN; HEADACHE, WEAKNESS, DIZZINESS, AND/OR LOSS OF APPETITE, HEAT CRAMPS, NAUSEA (WITH OR WITHOUT VOMITING), URGE TO DEFECATE, CHILLS (GOOSE-FLESH), RAPID BREATHING, CONFUSION, AND TINGLING OF THE HANDS AND/OR FEET. | 1. MOVE THE CASUALTY TO A COOL, SHADY AREA OR IMPROVISE SHADE AND LOOSEN OR REMOVE HIS CLOTHING.[2]<br><br>2. POUR WATER ON HIM AND FAN HIM TO PERMIT THE COOLANT EFFECT OF EVAPORATION.<br><br>3. HAVE HIM SLOWLY DRINK AT LEAST ONE CANTEEN FULL OF COOL WATER.<br><br>4. ELEVATE THE CASUALTY'S LEGS.<br><br>5. SEEK MEDICAL ASSISTANCE IF SYMPTOMS CONTINUE; MONITOR UNTIL SYMPTOMS ARE GONE OR MEDICAL ASSISTANCE ARRIVES. |
| **HEATSTROKE[3] (SUNSTROKE)** | THE CASUALTY STOPS SWEATING (RED [FLUSHED] HOT, DRY SKIN). HE FIRST MAY EXPERIENCE HEADACHE, DIZZINESS, NAUSEA, FAST PULSE AND RESPIRATION, SEIZURES, AND MENTAL CONFUSION. HE MAY COLLAPSE | 1. MOVE THE CASUALTY TO A COOL, SHADY AREA OR IMPROVISE SHADE AND LOOSEN OR REMOVE HIS CLOTHING, REMOVE THE OUTER GARMENTS AND PROTECTIVE |

*Table 5-1. Heat Injuries (Continued)*

| INJURIES | SIGNS AND SYMPTOMS | FIRST AID[1] |
|---|---|---|
| | AND SUDDENLY BECOME UNCONSCIOUS. **THIS IS A MEDICAL EMERGENCY.** | CLOTHING IF THE SITUATION PERMITS.[2]<br><br>2. START COOLING THE CASUALTY IMMEDIATELY. SPRAY OR POUR WATER ON HIM. FAN HIM. MASSAGE HIS EXTREMITIES AND SKIN.<br><br>3. ELEVATE HIS LEGS.<br><br>4. IF CONSCIOUS, HAVE HIM SLOWLY DRINK AT LEAST ONE CANTEEN FULL OF COOL WATER.<br><br>5. SEEK MEDICAL AID. CONTINUE COOLING WHILE AWAITING TRANSPORT AND CONTINUE FIRST AID WHILE EN ROUTE. |

**LEGEND**:

1 THE FIRST AID PROCEDURE FOR HEAT RELATED INJURIES CAUSED BY WEARING INDIVIDUAL PROTECTIVE EQUIPMENT (IPE) IS TO MOVE THE CASUALTY TO A CLEAN AREA AND GIVE HIM WATER TO DRINK.
2 WHEN IN A CHEMICAL ENVIRONMENT, DO NOT LOOSEN OR REMOVE THE CASUALTY'S CLOTHING.
3 CAN BE FATAL IF NOT PROVIDED FIRST AID AND MEDICAL TREATMENT PROMPTLY.

## 5-3. Cold Injuries

Cold injuries are most likely to occur when conditions are moderately cold, but accompanied by wet or windy conditions. Cold injuries can usually be prevented. Well-disciplined and well-trained service members can be protected even in the most adverse circumstances. They and their leaders must know the hazards of exposure to the cold. They must know the importance of personal hygiene, exercise, care of the feet and hands, and the use of protective clothing.

*a. Contributing Factors.*

(1) Temperature, humidity, precipitation, and wind greatly increase likelihood of cold injuries, and the service members with wet clothing are at great risk of cold injuries. Riverine operations (river, swamp, and stream crossings) increase likelihood of cold injuries. Low temperatures and low relative humidity (dry cold) promote frostbite. Higher temperatures, together with moisture, promote immersion syndrome. Windchill accelerates the loss of body heat and may aggravate cold injuries.

(2) Relatively stationary activities such as being in an observation post or on guard duty increase the service member's vulnerability to cold injury. Also, a service member is more likely to receive a cold injury if he is—

- In contact with the ground (such as marching, performing guard duty, or engaging in other outside activities).

- Immobile for long periods (such as while riding in an unheated or open vehicle).

- Standing in water, such as in a foxhole.

- Out in the cold for days without being warmed.

- Deprived of an adequate diet and rest.

- Not able to take care of his personal hygiene.

(3) Physical fatigue contributes to apathy, which leads to inactivity, personal neglect, carelessness, and reduced heat production. In turn, these increase the risk of cold injury. Service members with prior cold injuries have a higher-than-normal risk of subsequent cold injury; not necessarily involving the body part previously injured.

(4) Depressed or unresponsive service members are also vulnerable because they are less active. These service members tend to be careless about precautionary measures, especially warming activities, when cold injury is a threat.

(5) Excessive use of alcohol or drugs leading to faulty judgment or unconsciousness in a cold environment increases the risk of becoming a cold injury casualty.

*b.* *Signs and Symptoms.* Once a service member becomes familiar with the factors that contribute to cold injury, he must learn to recognize cold injury signs and symptoms.

(1) Many service members suffer cold injury without realizing what is happening to them. They may be cold and generally uncomfortable. These service members often do not notice the injured part because it is already numb from the cold.

(2) Superficial cold injury usually can be detected by numbness or tingling sensations. These signs and symptoms often can be relieved simply by loosening boots or other clothing and by exercising to improve circulation. In more advanced cases involving deep cold injury, the service member often is not aware that there is a problem until the affected part feels like a stump or block of wood.

(3) Outward signs of cold injury include discoloration of the skin at the site of injury. In light-skinned persons, the skin first reddens and then becomes pale or waxy white. In dark-skinned persons, grayness in the skin is usually evident. An injured foot or hand feels cold to the touch. Swelling may be an indication of deep injury. Also note that blisters may occur after rewarming the affected parts. Service members should work in pairs (buddy teams) to check each other for signs of discoloration and other symptoms.

*c. First Aid Measures.* First aid for cold injuries depends on whether they are superficial or deep. Rewarming the affected part using body heat can adequately treat cases of superficial cold injury. (For example, this can be done by covering cheeks with hands, putting fingertips in armpits, or placing the casualty's feet under the clothing of a buddy [next to his belly].) The injured part should **NOT** be massaged, exposed to a fire or stove, rubbed with snow, slapped, chafed, or soaked in cold water. Walking on injured feet should be avoided. Deep cold injury (frostbite) is very serious and requires prompt first aid to avoid or to minimize the loss of parts or all of the fingers, toes, hands, or feet. The sequence for treating cold injuries depends on whether the condition is life-threatening. The first priority in managing cold injuries is to remove the casualty from the cold environment (such as building an improvised shelter). Other injuries the casualty may have are provided first aid simultaneously while waiting for transportation or evacuation. If the casualty is to be transported in a nonmedical vehicle, first aid measures should be continued en route to the MTF.

*d. Conditions Caused by Cold.* Conditions caused by cold include chilblain, immersion syndrome (immersion foot and trench foot), frostbite, snow blindness, dehydration, and hypothermia.

(1) *Chilblain.*

- *Signs and symptoms.* Chilblain is caused by repeated prolonged exposure of bare skin at temperatures from 60° Fahrenheit (F) to

32°F, or 20°F for acclimated, dry, unwashed skin. The area may be acutely swollen, red, tender, and hot with itchy skin. There may be no loss of skin tissue in untreated cases but continued exposure may lead to infected, ulcerated, or bleeding lesions.

- *First aid measures.* Within minutes, the area usually responds to locally applied body heat. Rewarm the affected part by applying firm steady pressure with your hands, or placing the affected part under your arms or against the stomach of a buddy. **DO NOT** rub or massage affected areas.

**NOTE**

Medical personnel should evaluate the injury, because signs and symptoms of tissue damage may be slow to appear.

(2) *Immersion syndrome (immersion foot and trench foot).* Immersion foot and trench foot are injuries that result from fairly long exposure of the feet to wet conditions at temperatures from approximately 32°F to 50°F. Inactive feet in damp or wet socks and boots, or tightly laced boots which impair circulation, are even more susceptible to injury. This injury can be very serious; it can lead to loss of toes or parts of the feet. If exposure of the feet has been prolonged and severe, the feet may swell so much that pressure closes the blood vessels and cuts off circulation. Should an immersion injury occur, dry the feet thoroughly and transport the casualty to an MTF by the fastest means possible.

- *Signs and symptoms.* At first, the parts of the affected foot are cold and painless, the pulse is weak, and numbness may be present. Second, the parts may feel hot, and burning and shooting pains may begin. In later stages, the skin is pale with a bluish cast and the pulse decreases. Other signs and symptoms that may follow are blistering, swelling, redness, heat, hemorrhaging (bleeding), and gangrene.

- *First aid measures.* First aid measures are required for all stages of immersion syndrome injury. Rewarm the injured part gradually by exposing it to warm air. Protect it from trauma and secondary infections. Dry, loose clothing or several layers of warm coverings are preferable to extreme heat. Under no circumstances should the injured part be exposed to an open fire. Elevate the injured part to relieve the swelling. Transport the casualty to an MTF as soon as possible. When the part is rewarmed, the casualty often feels a burning sensation and pain. Symptoms may persist for days or weeks even after rewarming.

**NOTE**

When providing first aid for immersion foot and trench foot— **DO NOT** massage the injured part. **DO NOT** moisten the skin. **DO NOT** apply heat or ice.

(3) *Frostbite.* Frostbite is the injury of tissue caused from exposure to cold, usually below 32°F depending on the windchill factor, duration of exposure, and adequacy of protection. Individuals with a history of cold injury are likely to suffer an additional cold injury. The body parts most easily frostbitten are the cheeks, nose, ears, chin, forehead, wrists, hands, and feet. Frostbite may involve only the skin (superficial), or it may extend to a depth below the skin (deep). Deep frostbite is very serious and requires prompt first aid to avoid or to minimize the loss of parts or all of the fingers, toes, hands, or feet.

- *Signs and symptoms.*
  - Loss of sensation (numb feeling) in any part of the body.
  - Sudden blanching (whitening) of the skin of the affected part, followed by a momentary tingling sensation.
  - Redness of skin in light-skinned service members; grayish coloring in dark-skinned service members.
  - Blisters.
  - Swelling or tender areas.
  - Loss of previous sensation of pain in affected area.
  - Pale, yellowish, waxy-looking skin.
  - Frozen tissue that feels solid (or wooden) to the touch.

**CAUTION**

Deep frostbite is a very serious injury and requires immediate first aid and subsequent medical treatment to avoid or minimize loss of body parts.

- *First aid measures.*

  - *Face, ears, and nose.* Cover the casualty's affected area with his and/or your bare hands until sensation and color return.

  - *Hands.* Open the casualty's field jacket and shirt. (In a chemical environment, do not loosen or remove the clothing and protective overgarments.) Place the affected hands under the casualty's armpits. Close the field jacket and shirt to prevent additional exposure.

  - *Feet.* Remove the casualty's boots and socks if he does not need to walk any further to receive additional treatment. (Thawing the casualty's feet and forcing him to walk on them will cause additional pain and injury.) Place the affected feet under clothing and against the body of another service member.

**WARNING**

**DO NOT attempt to thaw the casualty's feet or other frozen areas if he will be required to walk or travel to an MTF for additional medical treatment. The possibility of additional injury from walking is less when the feet are frozen than when they are thawed. (However, if possible avoid walking.) Thawing in the field increases the possibilities of infection, gangrene, or other injury.**

**NOTE**

Thawing may occur spontaneously during transportation to the MTF; this cannot be avoided since the body in general must be kept warm.

In all of the above areas, ensure that the casualty is kept warm and that he is covered (to avoid further injury). Seek medical treatment as soon as possible. Reassure the casualty, protect the affected area from further injury by covering it lightly with a blanket or any dry clothing, and seek shelter out of the wind. Remove or loosen constricting clothing (except in a contaminated environment) and increase insulation. Ensure the casualty exercises as much as possible, avoiding trauma to the injured part, and is prepared for pain when thawing occurs. Protect the frostbitten part from additional injury. **DO NOT**—

- Rub the injured part with snow or apply cold water soaks.

- Warm the part by massage or exposure to open fire because the frozen part may be burned due to the lack of feeling.

- Use ointments or other salves.

- Manipulate the part in any way to increase circulation.

- Use alcohol or tobacco because this reduces the body's resistance to cold.

**NOTE**

Remember, when freezing extends to a depth below the skin, it is a much more serious injury. Extra care is required to reduce or avoid the chances of losing all or part of the toes or feet. This also applies to the fingers and hands.

(4) *Snow blindness.* Snow blindness is the effect that glare from an ice field or snowfield has on the eyes. It is more likely to occur in hazy, cloudy weather than when the sun is shining. Glare from the sun will cause an individual to instinctively protect his eyes. However, in cloudy weather, he may be overconfident and expose his eyes longer than when the threat is more obvious. He may also neglect precautions such as the use of protective eyewear. Waiting until discomfort (pain) is felt before using protective eyewear is dangerous because a deep burn of the eyes may already have occurred.

- *Signs and symptoms.* Symptoms of snow blindness are a sensation of grit in the eyes with pain in and over the eyes, made worse by moving the eyeball. Other signs and symptoms are watering, redness, headache, and increased pain on exposure to light.

- *First aid measures.* First aid measures consist of blindfolding or covering the eyes with a dark cloth which stops painful eye movement. Complete rest is desirable. If further exposure to light is not preventable, the eyes should be protected with dark bandages or the darkest glasses available. Once unprotected exposure to sunlight stops, the condition usually heals in a few days without permanent damage. The casualty should be evacuated to the nearest MTF.

(5) *Dehydration.* Dehydration occurs when the body loses too much fluid. A certain amount of body fluid is lost through normal body processes. A normal daily intake of liquids replaces these losses. When individuals are engaged in any strenuous exercises or activities, fluid is lost

through sweating and this loss creates an imbalance of fluids in the body, and if not matched by rehydration it can contribute to dehydration. The danger of dehydration is as prevalent in cold regions as it is in hot regions. In hot weather, the individual is aware of his body losing fluids through sweat. In cold weather, however, it is extremely difficult to realize that this condition exists since sweating is not as apparent as in a hot environment. The danger of dehydration in cold weather operations is a serious problem. In cold climates, sweat evaporates so rapidly or is absorbed so thoroughly by layers of heavy clothing that it is rarely visible on the skin. Dehydration also occurs during cold weather operations because drinking is inconvenient. Dehydration will weaken or incapacitate a casualty for a few hours, or sometimes, several days. Because rest is an important part of the recovery process, casualties must take care that limited movement during their recuperative period does not enhance the risk of becoming a cold injury casualty.

- *Signs and symptoms*. The symptoms of cold weather dehydration are similar to those encountered in heat exhaustion. The mouth, tongue, and throat become parched and dry, and swallowing becomes difficult. The casualty may have nausea (with or without vomiting) along with extreme dizziness and fainting. The casualty may also feel generally tired and weak and may experience muscle cramps. Focusing the eyes may also become difficult.

- *First aid measures*. The casualty should be kept warm and his clothes should be loosened (if not in a chemical environment) to allow proper circulation. Shelter from wind and cold must be provided. Fluid replacement should begin immediately and the service member transported to an MTF as soon as possible.

(6) *Hypothermia (general cooling)*. When exposed to prolonged cold weather a service member may become both mentally and physically numb, thus neglecting essential tasks or requiring more time and effort to achieve them. Under some conditions (particularly cold water immersion), even a service member in excellent physical condition may die in a matter of minutes. The destructive influence of cold on the body is called *hypothermia*. This means bodies lose heat faster than they can produce it. Hypothermia can occur from exposure to temperatures either above or below freezing, especially from immersion in cold water, wet-cold conditions, or from the effect of wind. Physical exhaustion and insufficient food intake may also increase the risk of hypothermia. General cooling of the entire body to a temperature below 95°F is caused by continued exposure to low or rapidly dropping temperatures, cold moisture, snow, or ice. Fatigue, poor physical condition, dehydration, faulty blood circulation, alcohol or other drug use, trauma, and immersion can cause hypothermia. Remember, cold

may affect the body systems slowly and almost without notice. Service members exposed to low temperatures for extended periods may suffer ill effects even if they are well protected by clothing.

- *Signs and symptoms*. As the body cools, there are several stages of progressive discomfort and impairment. A sign that is noticed immediately is shivering. Shivering is an attempt by the body to generate heat. The pulse is faint or very difficult to detect. People with temperatures around 90°F may be drowsy and mentally slow. Their ability to move may be hampered, stiff, and uncoordinated, but they may be able to function minimally. Their speech may be slurred. As the body temperature drops further, shock becomes evident as the person's eyes assume a glassy state, breathing becomes slow and shallow, and the pulse becomes weaker or absent. The person becomes very stiff and uncoordinated. Unconsciousness may follow quickly. As the body temperature drops even lower, the extremities freeze, and a deep (or core) body temperature (below 85°F) increases the risk of irregular heart action. This irregular heart action or heart standstill can result in sudden death.

- *First aid measures*. Except in cases of the most severe hypothermia (marked by coma or unconsciousness and a weak pulse), first aid measures for hypothermia are directed towards protecting the casualty from further loss of body heat. For the casualty who is conscious, first aid measures are directed at rewarming the body evenly and without delay. Provide heat by using a hot water bottle or field expedient or another service member's body heat.

**CAUTION**

**DO NOT** expose the casualty to an open fire, as he may become burned.

**NOTE**

When using a hot water bottle or field expedient (canteen filled with warm water), the bottle or canteen must be wrapped in cloth prior to placing it next to the casualty. This will reduce the chance of burning the casualty's skin.

Always call or send for help as soon as possible and protect the casualty immediately with dry clothing or a sleeping bag. Then, move him to a warm place. Evaluate other injuries and provide first aid as required. First aid measures can be performed while the casualty is waiting transportation or while he is en route. In the case of an accidental breakthrough into ice water, or other hypothermic accident, strip the casualty of wet clothing immediately and bundle him into a sleeping bag. Rescue breathing should be started at once if the casualty's breathing has stopped or is irregular or shallow. Warm liquids (**NOT HOT**) may be given gradually if the casualty is conscious. **DO NOT** force liquids on an unconscious or semiconscious casualty because he may choke. The casualty should be transported on a litter because the exertion of walking may aggravate circulation problems. Medical personnel should immediately treat any hypothermia casualty. Hypothermia is life threatening until normal body temperature has been restored. The first aid measures for a casualty with severe hypothermia are based upon the following principles: attempt to avoid further heat loss, handle the casualty gently, and transport the casualty as soon as possible to the nearest MTF. If at all possible, the casualty should be evacuated by medical personnel.

**WARNING**

**Rewarming a severely hypothermic casualty is extremely dangerous in the field due to the possibility of such complications as rewarming, shock and disturbances in the rhythm of the heartbeat. These conditions require treatment by medical personnel.**

NOTE

Resuscitation of casualties with hypothermic complications is difficult if not impossible to do outside of an MTF setting.

**CAUTION**

The casualty is unable to generate his own body heat. Therefore, merely placing him in a blanket or sleeping bag is not sufficient.

*e.* *Table*. See Table 5-2 for further information.

*Table 5-2. Injuries Caused by Cold and Wet Conditions*

| INJURIES | SIGNS/SYMPTOMS | FIRST AID |
|---|---|---|
| **CHILBLAIN** | RED SWOLLEN, HOT, TENDER, ITCHING SKIN. CONTINUED EXPOSURE MAY LEAD TO INFECTED (ULCERATED OR BLEEDING) SKIN LESIONS. | 1. AREA USUALLY RESPONDS TO LOCALLY APPLIED REWARMING (BODY HEAT).<br><br>2. DO NOT RUB OR MASSAGE AREA.<br><br>3. SEEK MEDICAL AID. |
| **IMMERSION SYNDROME (IMMERSION FOOT/TRENCH FOOT)** | AFFECTED PARTS ARE COLD, NUMB, AND PAINLESS. PARTS MAY THEN BE HOT, WITH BURNING AND SHOOTING PAINS. ADVANCED STAGE: SKIN PALE WITH BLUISH CAST; PULSE DECREASES; BLISTERING, SWELLING, HEAT, HEMORRHAGING, AND GANGRENE MAY FOLLOW. | 1. GRADUAL REWARMING BY EXPOSURE TO WARM AIR.<br><br>2. DO NOT MASSAGE OR MOISTEN SKIN.<br><br>3. PROTECT AFFECTED PARTS FROM TRAUMA.<br><br>4. DRY FEET THOROUGHLY, AVOID WALKING.<br><br>5. SEEK MEDICAL AID. |
| **FROSTBITE** | LOSS OF SENSATION (NUMB FEELING) IN ANY PART OF THE BODY. SUDDEN BLANCHING (WHITENING) OF THE SKIN OF THE AFFECTED PART, FOLLOWED BY A MOMENTARY TINGLING SENSATION. REDNESS OF SKIN IN LIGHT-SKINNED SERVICE MEMBERS; GRAYISH COLORING IN DARK-SKINNED SERVICE MEMBERS. BLISTERS. SWELLING OR TENDER AREAS. LOSS OF PREVIOUS SENSATION OF PAIN IN THE AFFECTED AREA. PALE YELLOWISH, WAXY-LOOKING SKIN. FROZEN TISSUE THAT FEELS SOLID (WOODEN) TO THE TOUCH. | 1. WARM THE AREA AT THE FIRST SIGN OF FROSTBITE, USING FIRM, STEADY PRESSURE OF THE HAND, UNDERARM, OR ABDOMEN.<br><br>2. FACE, EARS, NOSE: COVER AREA WITH HANDS (CASUALTY'S OWN OR BUDDY'S).<br><br>3. HANDS: OPEN FIELD JACKET AND PLACE CASUALTY'S HANDS AGAINST HIS BODY, THEN CLOSE THE JACKET TO PREVENT HEAT LOSS.<br><br>4. FEET: REMOVE THE CASUALTY'S BOOTS AND SOCKS AND PLACE HIS FEET AGAINST THE BODY OF ANOTHER SERVICE MEMBER. |

*Table 5-2. Injuries Caused by Cold and Wet Conditions (Continued)*

| INJURIES | SIGNS/SYMPTOMS | FIRST AID |
|---|---|---|
| | | **5. WARNING: DO NOT ATTEMPT TO THAW THE CASUALTY'S FEET OR OTHER FROZEN AREAS IF HE WILL BE REQUIRED TO WALK OR TRAVEL TO AN MTF FOR ADDITIONAL TREATMENT. THE POSSIBILITY OF INJURY FROM WALKING IS LESS WHEN THE FEET ARE FROZEN THAN WHEN THEY HAVE BEEN THAWED. (HOWEVER, IF POSSIBLE AVOID WALKING.) THAWING IN THE FIELD INCREASES THE POSSIBILITY OF INFECTION, GANGRENE, OR INJURY.**<br><br>6. LOOSEN OR REMOVE CONSTRICTING CLOTHING AND REMOVE ANY JEWELRY.<br><br>7. INCREASE INSULATION (COVER WITH BLANKET OR OTHER DRY MATERIAL). ENSURE CASUALTY EXERCISES AS MUCH AS POSSIBLE, AVOIDING TRAUMA TO INJURED PART. |
| **SNOW BLINDNESS** | EYES MAY FEEL SCRATCHY. WATERING, REDNESS, HEADACHE, AND INCREASED PAIN WITH EXPOSURE TO LIGHT CAN OCCUR. | 1. COVER THE EYES WITH A DARK CLOTH.<br><br>2. SEEK MEDICAL AID. |
| **DEHYDRATION** | SIMILAR TO HEAT EXHAUSTION (REFER TO TABLE 5-1). | 1. KEEP WARM.<br><br>2. CASUALTY NEEDS FLUID REPLACEMENT, REST, AND PROMPT MEDICAL AID. |
| **HYPOTHERMIA** | CASUALTY IS COLD. SHIVERING. CORE TEMPERATURE IS LOW. CONSCIOUSNESS MAY BE | **MILD HYPOTHERMIA**<br><br>1. REWARM BODY |

*Table 5-2. Injuries Caused by Cold and Wet Conditions (Continued)*

| INJURIES | SIGNS/SYMPTOMS | FIRST AID |
|---|---|---|
| | ALTERED. UNCOORDINATED MOVEMENTS MAY OCCUR. SHOCK AND COMA MAY RESULT AS BODY TEMPERATURE DROPS. | EVENLY AND WITHOUT DELAY. (NEED TO PROVIDE HEAT SOURCE; CASUALTY'S BODY UNABLE TO GENERATE HEAT.)<br><br>2. KEEP DRY, PROTECT FROM THE ELEMENTS.<br><br>3. WARM (NOT HOT) LIQUIDS MAY BE GIVEN GRADUALLY (TO CONSCIOUS CASUALTIES ONLY).<br><br>4. BE PREPARED TO START BASIC LIFE SUPPORT MEASURES FOR THE CASUALTY.<br><br>5. SEEK MEDICAL TREATMENT IMMEDIATELY.<br><br>**SEVERE HYPOTHERMIA**<br><br>1. STABILIZE THE TEMPERATURE.<br><br>2. ATTEMPT TO AVOID FURTHER HEAT LOSS.<br><br>3. HANDLE THE CASUALTY GENTLY.<br><br>4. EVACUATE TO THE NEAREST MTF AS SOON AS POSSIBLE.<br><br>**5. WARNING: HYPOTHERMIA IS A MEDICAL EMERGENCY. PROMPT MEDICAL ATTENTION IS NECESSARY.** |

CHAPTER 6

# FIRST AID FOR BITES AND STINGS

## 6-1. General

Snakebites, insect bites, or stings can cause intense pain and/or swelling. If not treated promptly and correctly, they can cause serious illness or death. The severity of a snakebite depends upon: whether the snake is poisonous or nonpoisonous, the type of snake, the location of the bite, and the amount of venom injected. Bites from humans and other animals, such as dogs, cats, bats, raccoons, and rats, can cause severe bruises and infection and tears or lacerations of tissue. Awareness of the potential sources of injuries can reduce or prevent them from occurring. Knowledge and prompt application of first-aid measures can lessen the severity of injuries from bites and stings and keep the service member from becoming a serious casualty.

## 6-2. Types of Snakes

*a. Nonpoisonous Snakes.* There are approximately 130 different varieties of nonpoisonous snakes in the United States. They have oval-shaped heads and round eyes. Unlike poisonous snakes, discussed below, nonpoisonous snakes do not have fangs with which to inject venom. Figure 6-1 depicts the characteristics of a nonpoisonous snake.

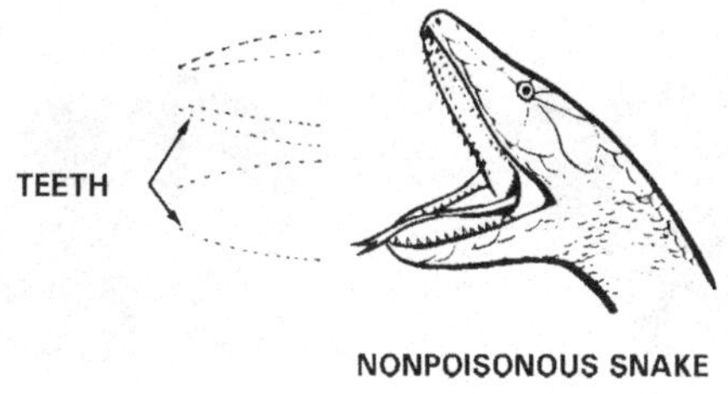


*Figure 6-1. Characteristics of nonpoisonous snake.*

*b. Poisonous Snakes.* Poisonous snakes are found throughout the world, primarily in tropical to moderate climates. Within the United States, there are four kinds: rattlesnakes, copperheads, water moccasins (cottonmouth), and coral snakes. Poisonous snakes in other parts of the world include sea snakes, the fer-de-lance, the bushmaster, and the tropical rattlesnake in tropical Central America; the Malayan pit viper in the tropical Far East; the cobra in Africa and Asia; the mamba (or black mamba) in central and southern Africa; and the krait in India and Southeast Asia. Refer to Figure 6-2 for characteristics of a poisonous pit viper.

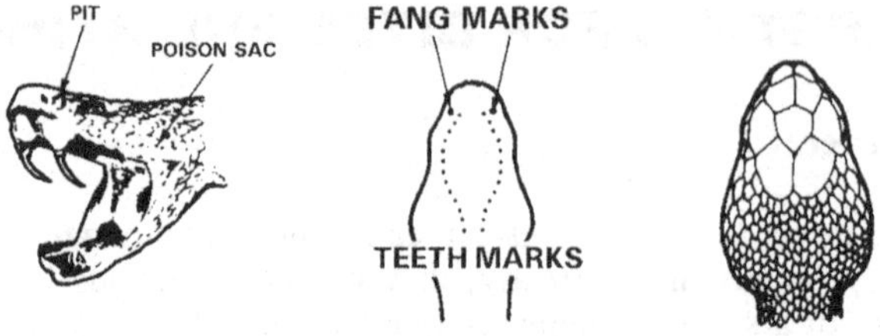


*Figure 6-2. Characteristics of poisonous pit viper.*

*c. Pit Vipers (Poisonous).* Figure 6-3 depicts a variety of poisonous snakes.

*Figure 6-3. Poisonous snakes.*

(1) Rattlesnakes, bushmasters, copperheads, fer-de-lance, Malayan pit vipers, and water moccasins (cottonmouth) are called pit vipers because of the small, deep pits between the nostrils and eyes on each side of the head (Figure 6-2). In addition to their long, hollow fangs, these snakes have other identifying features: thick bodies, slit-like pupils of the eyes, and flat, almost triangular-shaped heads. Color markings and other identifying characteristics, such as rattles or a noticeable white interior of the mouth (cottonmouth), also help distinguish these poisonous snakes. Further

identification is provided by examining the bite pattern of the wound for signs of fang entry. Occasionally there will be only one fang mark, as in the case of a bite on a finger or toe where there is no room for both fangs, or when the snake has broken off a fang.

(2) The casualty's condition provides the best information about the seriousness of the situation, or how much time has passed since the bite occurred. Pit viper bites are characterized by severe burning pain. Discoloration and swelling around the fang marks usually begins within 5 to 10 minutes after the bite. If only minimal swelling occurs within 30 minutes, the bite will almost certainly have been from a nonpoisonous snake or possibly from a poisonous snake which did not inject venom. The venom destroys blood cells, causing a general discoloration of the skin. Blisters and numbness in the affected area follow this reaction. Other signs, which can occur, are weakness, rapid pulse, nausea, shortness of breath, vomiting, and shock.

*d. Corals, Cobras, Kraits, and Mambas.* Corals (Figure 6-4), cobras (Figure 6-5), kraits, and mambas all belong to the same group even though they are found in different parts of the world. All four inject their venom through short, grooved fangs, leaving a characteristic bite pattern.

*Figure 6-4. Coral snake.*

(1) The small coral snake, found in the Southeastern US, is brightly colored with bands of red, yellow (or almost white), and black completely encircling the body. Other nonpoisonous snakes have the same coloring, but on the coral snake found in the US, the red ring always touches the yellow ring. To know the difference between a harmless snake and the coral snake found in the United States, remember the following:

"Red on yellow will kill a fellow,
Red on black, venom will lack."

*Figure 6-5. Cobra snake.*

(2) The venom of corals, cobras, kraits, and mambas produces symptoms different from those of pit vipers. Because there is only minimal pain and swelling, many people believe that the bite is not serious. Delayed reactions in the nervous system normally occur between 1 to 7 hours after the bite. Symptoms include blurred vision, drooping eyelids, slurred speech, drowsiness, and increased salivation and sweating. Nausea, vomiting, shock, respiratory difficulty, paralysis, convulsions, and coma will usually develop if the bite is not treated promptly.

*e. Sea Snakes.* Sea snakes (Figure 6-6) are found in the warm water areas of the Pacific and Indian oceans, along the coasts, and at the mouths of some larger rivers. Their venom is **VERY** poisonous, but their fangs are only 1/4 inch long. The first aid outlined for land snakes also applies to sea snakes.

*Figure 6-6. Sea snake.*

## 6-3. Snakebites

*a.* Poisonous snakes **DO NOT** always inject venom when they bite or strike a person. However, all snakes may carry tetanus (lockjaw); anyone bitten by a snake, whether poisonous or nonpoisonous, should immediately seek medical attention.

- Poison is injected from the venom sacs through grooved or hollow fangs. Depending on the species, these fangs are either long or short. Pit vipers have long hollow fangs. These fangs are folded against the roof of the mouth and extend when the snake strikes. This allows them to strike quickly and then withdraw. Cobras, coral snakes, kraits, mambas, and sea snakes have short, grooved fangs. These snakes are less effective in their attempts to bite, since they must chew after striking to inject enough venom (poison) to be effective. Figure 6-7 depicts the characteristics of a poisonous snakebite.

- In the event you are bitten, attempt to identify and/or kill the snake. Take it to medical personnel for inspection/identification. This provides valuable information to medical personnel who deal with snakebites. **TREAT ALL SNAKEBITES AS POISONOUS.**

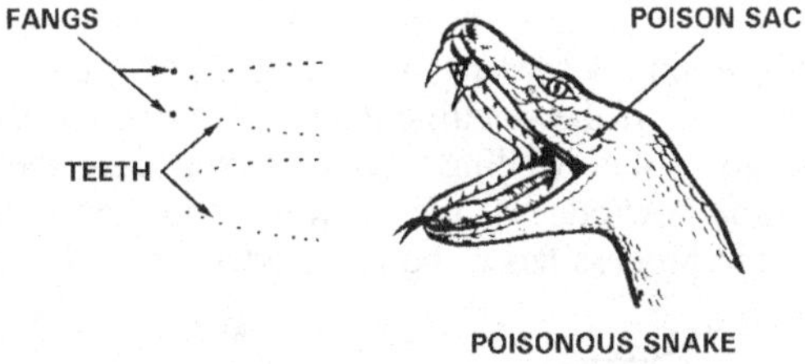


*Figure 6-7. Characteristics of poisonous snakebite.*

*b.* The venoms of different snakes cause different effects. Pit viper venom (hemotoxin [blood toxin]) destroys tissue and blood cells. Cobras, adders, and coral snakes inject powerful venom (neurotoxin [nerve toxin]) which affect the central nervous system, causing respiratory paralysis. Water moccasins and sea snakes have venom that is both hemotoxic and neurotoxic.

*c.* The identification of poisonous snakes is very important since medical treatment will be different for each type of venom. *Unless it can be positively identified, the snake should be killed and saved.* When this is not possible or when doing so is a serious threat to others, identification may sometimes be difficult since many venomous snakes resemble harmless varieties. When dealing with snakebite problems in foreign countries, seek advice, professional or otherwise, which may help identify species in the particular area of operations.

*d.* Get the casualty to an MTF as soon as possible and with minimum movement. Until evacuation or treatment is possible, have the casualty lie quietly and not move any more than necessary. If the casualty has been bitten on an extremity, **DO NOT** elevate the limb; keep the extremity level with the body. Keep the casualty comfortable and reassure him. If the casualty is alone when bitten, he should go to the medical facility himself rather than wait for someone to find him. Unless the snake has been positively identified, attempt to kill it and send it with the casualty. Be sure that retrieving the snake does not endanger anyone or delay transporting the casualty.

(1) If the bite is on an arm or leg, place a constricting band (narrow cravat [swathe], or narrow gauze bandage) one to two fingerbreadths above and below the bite (Figure 6-8). If the bite is on the hand or foot, place a single band above the wrist or ankle. The band should be tight enough to stop the flow of blood near the skin, but not tight enough to interfere with circulation. In other words, it should not have a tourniquet-like affect. If no swelling is seen, place the bands about 1 inch from either side of the bite. If swelling is present, put the bands on the unswollen part at the edge of the swelling. If the swelling extends beyond the band, move the band to the new edge of the swelling. (If possible, leave the old band on, place a new one at the new edge of the swelling, and then remove and save the old one in case the process has to be repeated.)

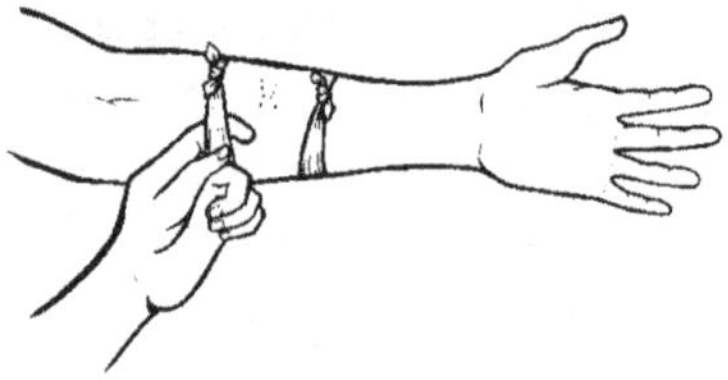

*Figure 6-8. Constricting band.*

**CAUTION**

**DO NOT** attempt to cut open the bite nor suck out the venom. If the venom should seep through any damaged or lacerated tissues in your mouth, you could immediately lose consciousness or even die.

(2) If the bite is located on an arm or leg, immobilize it at a level below the heart. **DO NOT** elevate an arm or leg even with or above the level of the heart.

**CAUTION**

When a splint is used to immobilize the arm or leg, take **EXTREME** care to ensure the splinting is done properly and does not bind. Watch it closely and adjust it if any changes in swelling occur.

(3) When possible, clean the area of the bite with soap and water. **DO NOT** use ointments of any kind.

(4) **NEVER** give the casualty food, alcohol, stimulants (coffee or tea), drugs, or tobacco.

(5) Remove rings, watches, or other jewelry from the affected limb.

## 6-4. Human or Animal Bites

Human or other land animal bites may cause lacerations or bruises. In addition to damaging tissue, bites always present the possibility of infection.

*a.* *Human Bites.* Human bites that break the skin may become seriously infected since the mouth is heavily contaminated with bacteria. Medical personnel **MUST** treat all human bites.

*b.* *Animal Bites.* Land animal bites can result in both infection and disease. Tetanus, rabies, and various types of fevers can follow an untreated animal bite. Because of these possible complications, the animal causing the bite should, if possible, be captured or killed (without damaging its head) so that it can be tested for disease.

*c.* *First Aid.*

(1) Cleanse the wound thoroughly with soap.

(2) Flush it well with water.

(3) Cover it with a sterile dressing.

(4) Immobilize the injured arm or leg, if appropriate.

(5) Transport the casualty immediately to an MTF.

NOTE

If unable to capture or kill the animal, provide medical personnel with any information that will help identify it.

## 6-5. Marine (Sea) Animals

With the exception of sharks and barracuda, most marine animals will not deliberately attack. The most frequent injuries from marine animals are wounds by biting, stinging, or puncturing. Wounds inflicted by marine animals can be very painful, but are rarely fatal.

*a. Sharks, Barracuda, and Alligators.* Wounds from these marine animals can involve major trauma as a result of bites and lacerations. Bites from large marine animals are potentially the most life threatening of all injuries from marine animals. Major wounds from these animals can be treated by controlling the bleeding, preventing shock, giving basic life support, splinting the injury, and by securing prompt medical aid.

*b. Turtles, Moray Eels, and Corals.* These animals normally inflict minor wounds. Treat by cleansing the wound(s) thoroughly and by splinting if necessary.

*c. Jellyfish, Portuguese Man-of-War, Anemones, and Others.* This group of marine animals inflict injury by means of stinging cells in their tentacles. Contact with the tentacles produces burning pain with a rash and small hemorrhages on the skin. Shock, muscular cramping, nausea, vomiting, and respiratory distress may also occur. Gently remove the clinging tentacles with a towel and wash or treat the area. Use diluted ammonia or alcohol, meat tenderizer, and talcum powder. If symptoms become severe or persist, seek medical assistance.

*d. Spiny Fish, Urchins, Stingrays, and Cone Shells.* These animals inject their venom by puncturing the skin with their spines. General signs and symptoms include swelling, nausea, vomiting, generalized cramps, diarrhea, muscular paralysis, and shock. Deaths are rare. Treatment consists of soaking the wounds in hot water (when available) for 30 to 60 minutes. This inactivates the heat sensitive toxin. In addition, further first aid measures (controlling bleeding, applying a dressing, and so forth) should be carried out as necessary.

**CAUTION**

Be careful not to scald the casualty with water that is too hot because the pain of the wound will mask the normal reaction to heat.

## 6-6. Insect (Arthropod) Bites and Stings

An insect bite or sting can cause great pain, allergic reaction, inflammation, and infection. If not treated correctly, some bites/stings may cause serious illness or even death. When an allergic reaction is not involved, first aid is a simple process. In any case, medical personnel should examine the casualty at the earliest possible time. It is important to properly identify the spider, bee, or creature that caused the bite/sting, especially in cases of allergic reaction.

*a. Types of Insects.* The insects found throughout the world that can produce a bite or sting are too numerous to mention in detail. Commonly encountered stinging or biting insects include brown recluse spiders (Figure 6-9), black widow spiders (Figure 6-10), tarantulas (Figure 6-11), scorpions (Figure 6-12), urticating caterpillars, bees, wasps, centipedes, conenose beetles (kissing bugs), ants, and wheel bugs. Upon being reassigned, especially to overseas areas, take the time to become acquainted with the types of insects to avoid.

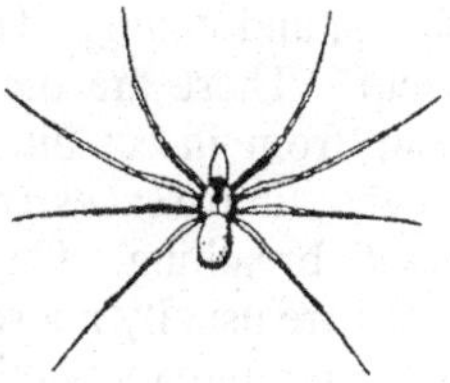

*Figure 6-9. Brown recluse spider.*

*Figure 6-10. Black widow spider.*

*Figure 6-11. Tarantula.*

*Figure 6-12. Scorpion.*

*b. Signs and Symptoms.* Discussed in paragraphs (1) and (2) below are the most common effects of insect bites/stings. They can occur alone or in combination with the others.

(1) *Less serious.* Commonly seen signs/symptoms are pain, irritation, swelling, heat, redness, and itching. Hives or wheals (raised areas of the skin that itch) may occur. These are the least severe of the allergic reactions that commonly occur from insect bites/stings. They are usually dangerous only if they affect the air passages (mouth, throat, nose, and so forth), which could interfere with breathing. The bites/stings of bees, wasps, ants, mosquitoes, fleas, and ticks are usually not serious and normally produce mild and localized symptoms. A tarantula's bite is usually no worse than that of a bee sting. Scorpions are rare and their stings (except for a specific species found only in the Southwest desert) are painful but usually not dangerous.

(2) *Serious.* Emergency allergic or hypersensitive reactions sometimes result from the stings of bees, wasps, and ants. Many people are allergic to the venom of these particular insects. Bites or stings from these insects may produce more serious reactions, to include generalized itching and hives, weakness, anxiety, headache, breathing difficulties, nausea, vomiting, and diarrhea. Very serious allergic reactions (called *anaphylactic shock*) can lead to complete collapse, shock, and even death. Spider bites (particularly from the black widow and brown recluse spiders) can also be

serious. Venom from the black widow spider affects the nervous system. This venom can cause muscle cramps, a rigid, nontender abdomen, breathing difficulties, sweating, nausea, and vomiting. The brown recluse spider generally produces local rather than system-wide problems; however, local tissue damage around the bite can be severe and can lead to an ulcer and even gangrene.

c. *First Aid.* There are certain principles that apply regardless of what caused the bite/sting. Some of these are—

- If there is a stinger present (for example, from a bee), remove the stinger by scraping the skin's surface with a fingernail or knife. **DO NOT** squeeze the sac attached to the stinger because it may inject more venom.

- Wash the area of the bite/sting with soap and water (alcohol or an antiseptic may also be used) to help reduce the chances of an infection and remove traces of venom.

- Remove jewelry from bitten extremities because swelling may occur.

- In most cases of insect bites the reaction will be mild and localized; use ice or cold compresses (if available) on the site of the bite/sting. This will help reduce swelling, ease the pain, and slow the absorption of venom. Meat tenderizer (to neutralize the venom) or calamine lotion (to reduce itching) may be applied locally. If necessary, seek medical assistance.

- In more serious reactions (severe and rapid swelling, allergic symptoms, and so forth) treat the bite/sting like you would treat a snakebite; that is, apply constricting bands above and below the site.

- Be prepared to perform basic life-support measures, such as rescue breathing.

- Reassure the casualty and keep him calm.

- In serious reactions, attempt to capture the insect for positive identification; however, be careful not to become a casualty yourself.

- If the reaction to the bite/sting appears serious, seek medical assistance.

**WARNING**

**Insect bites/stings may cause *anaphylactic shock* (a shock caused by a severe allergic reaction). This is a life-threatening event and a TRUE MEDICAL EMERGENCY. Be prepared to perform the basic life-support measures and to immediately transport the casualty to an MTF.**

NOTE

Be aware that some allergic or hypersensitive individuals may carry identification or emergency insect bite treatment kits. If the casualty is having an allergic reaction and has such a kit, administer the medication in the kit according to the instructions which accompany the kit.

*d. Supplemental Information.* For additional information concerning biting insects, see FM 21-10.

## 6-7. First Aid for Bites and Stings

See the table below for information on bites and stings.

*Table 6-1. First Aid Measures for Bites and Stings*

| TYPES | FIRST AID MEASURES |
|---|---|
| SNAKEBITE | 1. MOVE CASUALTY AWAY FROM THE SNAKE.<br>2. REMOVE JEWELRY FROM THE AFFECTED AREA, IF APPLICABLE.<br>3. REASSURE CASUALTY AND KEEP HIM QUIET.<br>4. APPLY CONSTRICTING BAND, 1-2 FINGERBREADTHS FROM THE BITE. YOU SHOULD BE ABLE TO INSERT A FINGER BETWEEN THE BAND AND THE SKIN.<br>*a. ARM OR LEG BITE.* PLACE ONE BAND ABOVE AND ONE BAND BELOW THE BITE SITE.<br>*b. HAND OR FOOT BITE.* PLACE ONE BAND ABOVE THE WRIST OR ANKLE. |

*Table 6-1. First Aid Measures for Bites and Stings*

| TYPES | FIRST AID MEASURES |
|---|---|
| | 5. IMMOBILIZE THE AFFECTED PART IN A POSITION BELOW THE LEVEL OF THE HEART.<br>6. KILL THE SNAKE (IF POSSIBLE, WITHOUT DAMAGING ITS HEAD OR ENDANGERING YOURSELF) AND SEND IT WITH THE CASUALTY.<br>7. SEEK MEDICAL ASSISTANCE IMMEDIATELY. |
| **BROWN RECLUSE SPIDER OR BLACK WIDOW SPIDER BITE** | 1. KEEP CASUALTY QUIET.<br>2. REMOVE ALL JEWELRY FROM AFFECTED PART, IF APPLICABLE.<br>3. WASH THE AREA.<br>4. APPLY ICE OR FREEZE PACK, IF AVAILABLE.<br>5. SEEK MEDICAL ASSISTANCE. |
| **TARANTULA BITE OR SCORPION STING OR ANT BITE** | 1. WASH THE AREA.<br>2. REMOVE ALL JEWELRY FROM AFFECTED PART, IF APPLICABLE.<br>3. APPLY ICE OR FREEZE PACK, IF AVAILABLE.<br>4. APPLY BAKING SODA, CALAMINE LOTION, OR MEAT TENDERIZER (IF AVAILABLE) TO BITE SITE TO RELIEVE PAIN AND ITCHING.<br>5. IF THE SITE OF THE BITE IS ON THE FACE, NECK (POSSIBLE AIRWAY PROBLEMS), OR GENITAL AREA, OR IF LOCAL REACTION SEEMS SEVERE, OR IF THE STING IS BY THE DANGEROUS TYPE OF SCORPION FOUND IN THE SOUTHWEST UNITED STATES DESERT, KEEP THE CASUALTY AS QUIET AS POSSIBLE. SEEK MEDICAL ASSISTANCE. |
| **BEE STING** | 1. IF THE STINGER IS PRESENT, REMOVE BY SCRAPING WITH A KNIFE OR FINGERNAIL. DO NOT SQUEEZE VENOM SAC ON STINGER; MORE VENOM MAY BE INJECTED.<br>2. REMOVE ALL JEWELRY FROM AFFECTED PART, IF APPLICABLE. |

*Table 6-1. First Aid Measures for Bites and Stings*

| TYPES | FIRST AID MEASURES |
|---|---|
| | 3. WASH THE AREA.<br>4. APPLY ICE OR FREEZE PACK, IF AVAILABLE.<br>5. IF ALLERGIC SIGNS OR SYMPTOMS APPEAR, BE PREPARED TO PERFORM BASIC LIFE SUPPORT MEASURES. SEEK IMMEDIATE MEDICAL ASSISTANCE. |

CHAPTER 7

# FIRST AID IN A NUCLEAR, BIOLOGICAL, AND CHEMICAL ENVIRONMENT

## 7-1. General

American forces have not been exposed to NBC weapons/agents on the battlefield since World War I. In future conflicts and wars we can expect the use of such agents. Nuclear, biological, and chemical weapons will rapidly degrade unit effectiveness by forcing troops to wear protective clothing and by creating confusion and fear. Through training in protective procedures and first aid, units can maintain their effectiveness on the integrated battlefield.

## 7-2. First Aid Materials

You may be issued the following materials to protect, decontaminate, and use as first aid for NBC exposure. You must know how to use the items; some items are described in *a* through *d* below. It is equally important that you know when to use them.

*a. Nerve Agent Pyridostigmine Pretreatment (NAPP).* You may be issued a blister pack of pretreatment tablets when your commander directs. The NAPP is a pretreatment; it is not an antidote. It improves the effectiveness of the nerve agent antidote. When ordered to take the pretreatment you must take one tablet every 8 hours, mission permitting. This must be taken prior to exposure to nerve agents, since it may take several hours to develop adequate blood levels.

**NOTE**

Commanders must follow investigational new drug protocols for use of the NAPP.

*b. M291 Skin Decontaminating Kit.* The M291 Skin Decontaminating Kit (Figure 7-1) contains six packets of XE-555 decontaminant resin.

**WARNING**

**For external use only. May be slightly irritating to the eyes. Keep decontaminating powder out of eyes. Use water to wash toxic agent out of eyes.**

*c. Nerve Agent Antidote Kit, MARK I.* Each service member is issued three MARK Is for use in first aid for nerve agent poisoning (Figure 7-2 and paragraph 7-6).

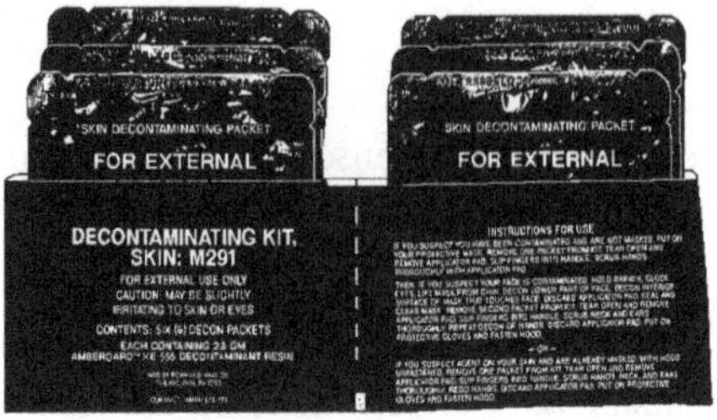


*Figure 7-1. M291 Skin Decontamination Kit.*

*d. Antidote Treatment, Nerve Agent, Autoinjector.* A new nerve agent antidote injection device, Antidote Treatment, Nerve Agent, Autoinjector (ATNAA) is currently under development that will replace the MARK I. The ATNAA is a multichambered device with the atropine and pralidoxime chloride in separate chambers. Both antidotes will be administered through a single needle.

## 7-3. Classification of Chemical and Biological Agents

*a.* Chemical agents are classified according to the primary physiological effects they produce, such as blistering, choking, vomiting, and incapacitating agents.

*b.* Biological warfare agents are classified according to the effect they have on man. The effects include their ability to incapacitate and cause death. Most biological warfare agents are delivered as aerosols that effect the respiratory tract; some can be delivered by releasing infected insects, by contaminating food and water, and by injection (injecting material in individuals by terrorist, not mass exposure). These agents are found in living organisms such as fungi, bacteria, and viruses.

**WARNING**

**Swallowing water or food contaminated with nerve, blister, and other chemical agents and with some biological agents can be fatal. NEVER consume water or food that is suspected of being contaminated until it has been tested and found safe for consumption by medical personnel.**

## 7-4. Conditions for Masking Without Order or Alarm

*a.* Once an attack with a chemical or biological agent is detected or suspected, or information is available that such an agent is about to be used, you must **STOP BREATHING** and mask immediately. **DO NOT WAIT** to receive an order or alarm under the following circumstances:

- Your position is hit by artillery missiles, rockets that produce vapors, smoke, and mists, and aerial sprays.
- Smoke or vapor cloud from an unknown source is present or approaching.
- A suspicious odor, liquid, or solid is present.
- A chemical or biological warfare agent attack is occurring.
- You are entering an area known or suspected of being contaminated.
- When casualties are being received from an area where chemical or biological agents have reportedly been used.
- You have one or more of the following symptoms:
  - An unexplained runny nose.
  - A sudden unexplained headache.
  - A feeling of choking or tightness in the chest or throat.
  - Dimness of vision.
  - Irritation of the eyes.
  - Difficulty in or increased rate of breathing without obvious reasons.
  - Sudden feeling of depression.
  - Dread, anxiety, or restlessness.
  - Dizziness or light-headedness.

- Slurred speech.
- Unexplained laughter or unusual behavior is noted in others.
- Numerous unexplained ill personnel.
- Service members suddenly collapsing without evident cause.
- Animals or birds exhibiting unusual behavior or suddenly dying.

*b.* For further information on protection and masking procedures, refer to FM 3-4, FM 4-02.7, FM 8-284, and FM 8-285.

## 7-5. First Aid for a Chemical Attack

Your field protective mask gives protection against biological and chemical warfare agents as well as radiological fallout. With practice you can mask in 9 seconds or less, or put on your mask with hood within 15 seconds.

*a.* Stop breathing. Don your mask, seal it properly, and clear and check it; then resume breathing. Give the alarm, and continue the mission. Keep your mask on until the "all clear" signal has been given.

**NOTE**

Keep your mask on until the area is no longer hazardous and you are told to unmask.

*b.* If symptoms of nerve agent poisoning (paragraph 7-7) appear, immediately give yourself one MARK I or ATNAA.

**CAUTION**

Do not inject a nerve agent antidote until you are sure you need it.

*c.* If your eyes and face become contaminated, you must immediately try to get under cover. You need shelter to prevent further contamination while performing decontamination procedures on your face. If no overhead cover is available, put your poncho over your head before beginning the decontamination process. Then you put on the remaining protective clothing. If vomiting occurs, the mask should be lifted momentarily and drained—with your eyes closed and while holding your breath—then replaced, cleared, and sealed.

*d.* If nerve agents are used, mission permitting, watch for persons needing nerve agent antidotes and immediately follow procedures outlined in paragraph 7-8*b* or *c*.

*e.* Decontaminate your skin immediately and clothing and equipment as soon as the mission permits.

## 7-6. Background Information on Nerve Agents

*a.* Nerve agents are among the deadliest of chemical agents. Nerve agents enter the body by inhalation, by ingestion, and through the skin. Depending on the route of entry and the amount, nerve agents can produce injury or death within minutes. Nerve agents can achieve their effects with small amounts. Nerve agents are absorbed rapidly, and the effects are felt immediately upon entry into the body. You will be issued three MARK Is or three ATNAAs and one Convulsant Antidote for Nerve Agent (CANA). Each MARK I consists of one atropine autoinjector and one pralidoxime chloride (2 PAM Cl) autoinjector (Figure 7-2A). Each ATNAA consist of a multichambered autoinjector with the atropine and pralidoxime chloride in separate chambers (Figure 7-2C). The CANA is a single autoinjector with flanges (Figure 7-2B). Procedures for use of both the MARK I and ATNAA are described below. You will use either the MARK I or the ATNAA in self-aid and buddy aid as issued.

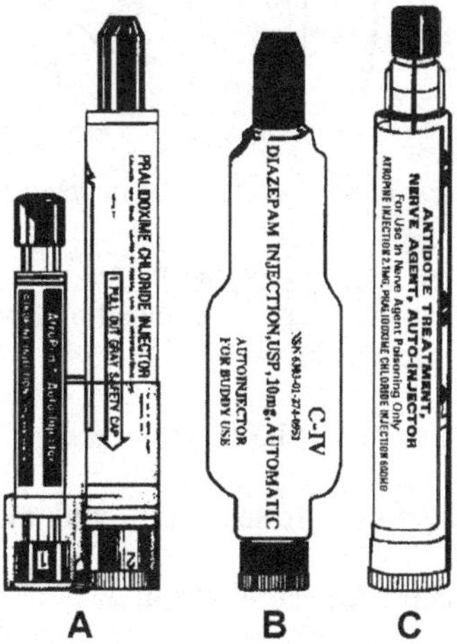


*Figure 7-2. Nerve Agent Antidote Kit, MARK I, CANA, and ATNAA.*

*b*. When you have the signs and symptoms of nerve agent poisoning, you should immediately put on the protective mask and then inject yourself with one set of the MARK I or ATNAA. Do not administer the CANA. You should inject yourself in the outer (lateral) thigh muscle (Figure 7-3) or if you are thin, in the upper outer (lateral) part of the buttocks (Figure 7-4).

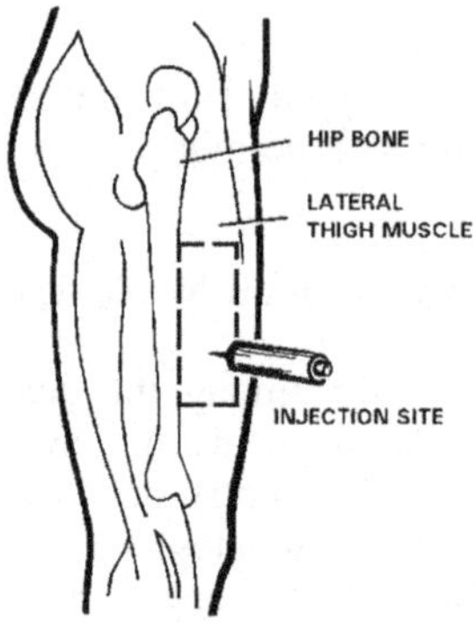


*Figure 7-3. Thigh injection site.*

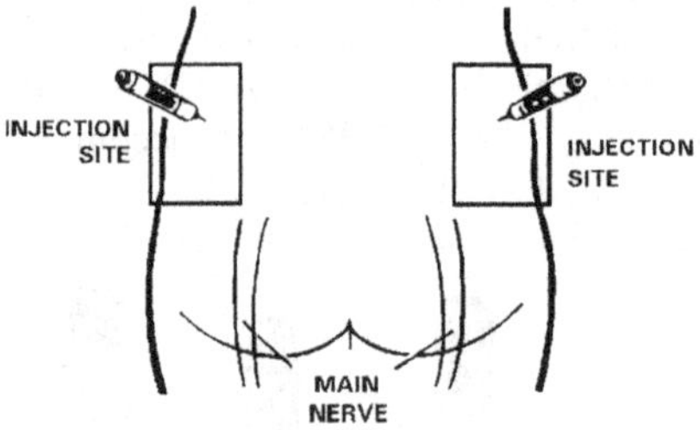


*Figure 7-4. Buttocks injection site.*

*c*. Also, you may come upon an unconscious chemical agent casualty who will be unable to care for himself and who will require first aid. You should be able to successfully—

(1) Mask him if he is unmasked.

(2) Inject him, if necessary, with all of **HIS** autoinjectors.

(3) Decontaminate his skin.

(4) Seek medical assistance.

## 7-7. Signs and Symptoms of Nerve Agent Poisoning

The symptoms of nerve agent poisoning are grouped as **MILD**—those that you recognize and for which you can perform self-aid, and **SEVERE**—those which require buddy aid.

*a. MILD Signs and Symptoms.*

- Unexplained runny nose.
- Unexplained sudden headache.
- Sudden drooling.
- Difficulty seeing (dimness of vision and miosis).
- Tightness in the chest or difficulty in breathing.
- Localized sweating and muscular twitching in the area of contaminated skin.
- Stomach cramps.
- Nausea.
- Tachycardia followed by bradycardia. (*Tachycardia* is an abnormally rapid heartbeat with a heart rate of over 100 beats per minute. *Bradycardia* is a slow heart rate of less than 60 beats per minute.)

*b. SEVERE Signs and Symptoms.*

- Strange or confused behavior.
- Wheezing, dyspnea (difficulty in breathing), and coughing.
- Severely pinpointed pupils.
- Red eyes with tearing.
- Vomiting.
- Severe muscular twitching and general weakness.
- Involuntary urination and defecation.

- Convulsions.
- Unconsciousness.
- Respiratory failure.
- Bradycardia.

## 7-8. First Aid for Nerve Agent Poisoning

First aid for nerve agent poisoning consists of administering the MARK I or ATNAA and CANA.

*a. Injection Site.* The injection site for administering the antidotes is normally in the outer thigh muscle. The thigh injection site is the area about a hand's width above the knee to a hand's width below the hip joint (Figure 7-3). It is important that the injection be given into a large muscle area. If the individual is thinly built, then the injections should be administered into the upper outer quarter (quadrant) of the buttock (Figure 7-4). Injecting in the buttocks of a thinly built individual avoids injury to the thighbone.

*b. Self-Administer MARK I.* If you experience any or all of the nerve agent **MILD** symptoms (paragraph 7-7*a*), you must **IMMEDIATELY** put on your protective mask and self-administer one MARK I (Figure 7-2A). Follow the procedure given in Table 7-1. The MARK I is carried in your protective mask carrier, pocket of the MOPP overgarment, or other location as specified in your unit tactical standing operating procedure (TSOP). (In cold weather, the MARK I should be stored in an inside pocket of your clothing to protect the antidote from freezing. A frozen MARK I cannot be immediately used to provide you with antidote, when needed. (However, the MARK I can still be used after complete thawing.)

*Table 7-1. Self Aid for Nerve Agent Poisoning*

| MARK I* | ATNAA* |
|---|---|
| STEP 1. OBTAIN ONE MARK I.** | STEP 1. OBTAIN ONE ATNAA.** |
| STEP 2. CHECK INJECTION SITE. | STEP 2. CHECK INJECTION SITE. |
| STEP 3. HOLD MARK I AT EYE LEVEL WITH NONDOMINANT HAND WITH THE LARGE INJECTOR ON TOP (FIGURE 7-5A). | STEP 3. HOLD ATNAA WITH DOMINANT HAND (FIGURE 7-12A). |

*Table 7-1. Self Aid for Nerve Agent Poisoning (Continued)*

| MARK I* | ATNAA* |
|---|---|
| STEP 4. GRASP SMALL INJECTOR (ATROPINE) (FIGURE 7-5B) AND REMOVE FROM CLIP (FIGURE 7-5C). | STEP 4. GRASP SAFETY CAP WITH NONDOMINANT HAND AND REMOVE FROM INJECTOR (FIGURE 7-12B). |
| STEP 5. CLEAR HARD OBJECTS FROM INJECTION SITE. | STEP 5. CLEAR HARD OBJECTS FROM INJECTION SITE. |
| STEP 6. INJECT ATROPINE AT INJECTION SITE APPLYING EVEN PRESSURE TO THE INJECTOR (FIGURE 7-6 OR 7-7). HOLD IN PLACE FOR 10 SECONDS. | STEP 6. INJECT ATNAA AT INJECTION SITE APPLYING EVEN PRESSURE TO THE INJECTOR (FIGURE 7-14 OR 7-15). HOLD IN PLACE FOR 10 SECONDS. |
| STEP 7. HOLD USED INJECTOR WITH NONDOMINANT HAND. | STEP 7. BEND NEEDLE OF USED INJECTOR BY PRESSING ON A HARD SURFACE TO FORM A HOOK. |
| STEP 8. GRASP THE LARGE (2 PAM CI) INJECTOR (FIGURE 7-8B) AND PULL IT FROM CLIP (FIGURE 7-8C). DROP CLIP TO GROUND. | STEP 8. ATTACH USED INJECTOR TO BLOUSE POCKET FLAP OF BDO/JSLIST (FIGURE 7-16). |
| STEP 9. INJECT 2 PAM CI AT INJECTION SITE APPLYING EVEN PRESSURE TO THE INJECTOR (FIGURE 7-6 OR 7-7). HOLD IN PLACE FOR 10 SECONDS. | STEP 9. MASSAGE INJECTION SITE, MISSION PERMITTING. |
| STEP 10. BEND THE NEEDLES OF ALL USED INJECTORS BY PRESSING ON A HARD SURFACE TO FORM A HOOK. | |
| STEP 11. ATTACH ALL USED INJECTORS TO BLOUSE POCKET FLAP OF BDO/JSLIST (FIGURE 7-9). | |
| STEP 12. MASSAGE INJECTION SITE, MISSION PERMITTING. | |

* USE STEPS LISTED FOR TYPE OF ANTIDOTE DEVICE ISSUED.
** ONLY ADMINISTER ONE MARK I OR ATNAA AS SELF-AID. DO NOT SELF-ADMINISTER CANA.

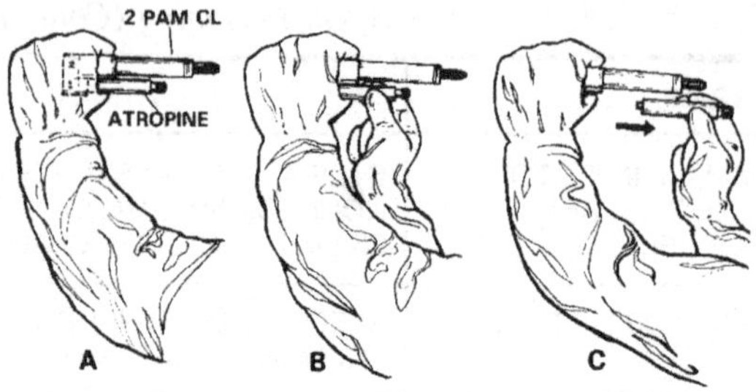


*Figure 7-5. Removing the atropine autoinjector from the MARK I clip.*

**CAUTION**

**DO NOT** cover or hold the needle end with your hand, thumb, or fingers—you might accidentally inject yourself. An accidental injection into the hand **WILL NOT** deliver an effective dose of the antidote, especially if the needle goes through the hand.

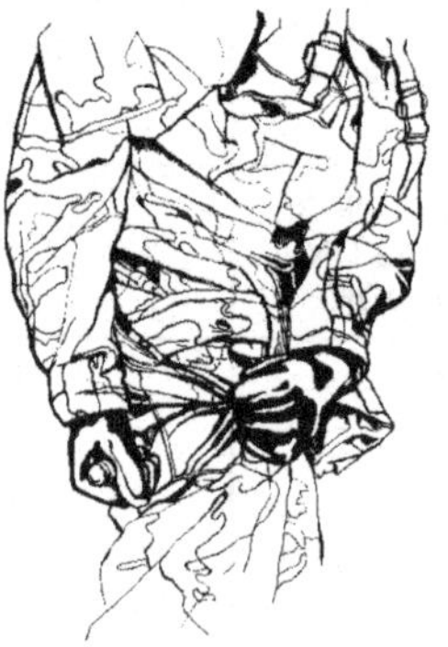

*Figure 7-6. Thigh injection site for self-aid.*

**NOTE**

If you are thinly built, inject yourself into the upper outer quadrant of the buttock (Figure 7-7). There is a nerve that crosses the buttocks; hitting this nerve can cause paralysis. Therefore, you must only inject into the *upper outer quadrant* of the buttock.

*Figure 7-7. Buttocks injection site for self-aid.*

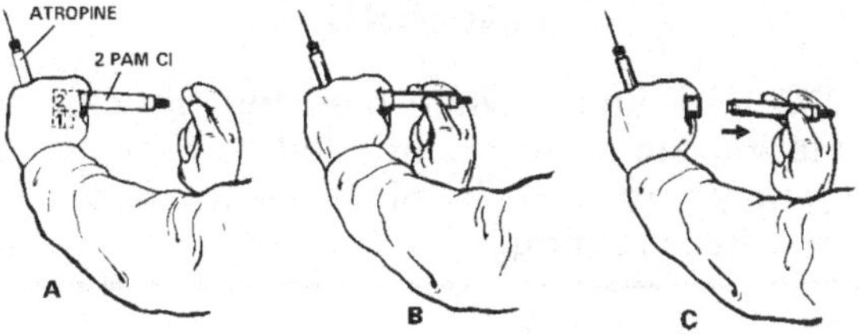


*Figure 7-8. Removing the 2 PAM Cl autoinjector from the MARK I clip.*

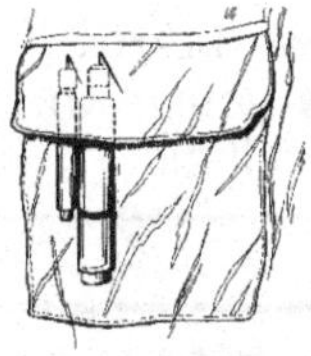

*Figure 7-9. One set of used MARK I autoinjectors attached to pocket flap.*

**NOTES**

1. **DO NOT** give yourself another set of injections. If you are able to walk without assistance, know who you are, and where you are, you **WILL NOT** need the second set of injections. (If not needed, giving yourself a second set of MARK I injections or ATNAA may create a nerve agent antidote overdose, which could cause incapacitation [inability to perform mission or defend yourself].)

2. If you continue to have symptoms of nerve agent poisoning, seek someone else (a buddy) to check your symptoms and administer the additional sets of injections, if required.

*c. Buddy Evaluation and Buddy Aid.* Service members may seek assistance after self-aid (self-administering one MARK I or ATNAA) or may become incapacitated after self-aid. A buddy must evaluate the individual to determine if additional antidotes are required to counter the effects of the nerve agent. Also, service members may experience **SEVERE** symptoms of nerve agent poisoning (paragraph 7-7*b*); they will not be able to treat themselves. In either case, other service members must perform buddy aid as quickly as possible. Before initiating buddy aid, determine if one set of MARK I autoinjectors has already been used so that no more than three sets of the antidote are administered. Buddy aid also includes administering the CANA with the third MARK I or ATNAA to prevent convulsions. Follow the procedures indicated in Table 7-2.

**WARNING**

**Squat, DO NOT kneel, when masking the casualty or administering the nerve agent antidote to the casualty. Kneeling may force the chemical agent into or through your protective clothing.**

**CAUTION**

**DO NOT** use your own MARK I, ATNAA, or CANA on a casualty. If you use your own, you may not have any antidote if needed for self-aid.

**WARNING**

**DO NOT inject into areas close to the hip, knee, or thigh-bone.**

*Table 7-2. Buddy Aid/Combat Lifesaver Aid for Nerve Agent Casualty.*

| MARK I* | ATNAA* | CANA** |
|---|---|---|
| STEP 1. MASK THE CASUALTY AND POSITION HIM ON HIS SIDE (SWIMMER'S POSITION). | STEP 1. MASK THE CASUALTY AND POSITION HIM ON HIS SIDE (SWIMMER'S POSITION). | STEP 1. OBTAIN BUDDY'S CANA. |
| STEP 2. POSITION YOURSELF NEAR THE CASUALTY'S THIGH. | STEP 2. POSITION YOURSELF NEAR THE CASUALTY'S THIGH. | STEP 2. CHECK INJECTION SITE. |

*Table 7-2. Buddy Aid/Combat Lifesaver Aid for Nerve Agent Casualty (Continued).*

| MARK I* | ATNAA* | CANA** |
|---|---|---|
| STEP 3. OBTAIN BUDDY'S THREE OR REMAINING MARK Is. | STEP 3. OBTAIN BUDDY'S THREE OR REMAINING ATNAAs. | STEP 3. HOLD CANA IN A CLOSED FIST WITH DOMINANT HAND (FIGURE 7-12A). |
| STEP 4. CHECK INJEC-TION SITE. | STEP 4. CHECK INJEC-TION SITE. | STEP 4. GRASP SAFETY CAP WITH NONDOMINANT HAND AND REMOVE FROM INJECTOR (FIGURE 7-12B). |
| STEP 5. HOLD MARK I WITH NONDOMINANT HAND (FIGURE 7-5A). | STEP 5. HOLD ATNAA IN A CLOSED FIST WITH DOMINANT HAND (FIGURE 7-12A). | STEP 5. CLEAR HARD OBJECTS FROM INJECTION SITE. |
| STEP 6. GRASP SMALL INJECTOR (ATROPINE) AND REMOVE FROM CLIP (FIGURE 7-5B). | STEP 6. GRASP SAFETY CAP WITH NONDOMINANT HAND AND REMOVE FROM INJECTOR (FIGURE 7-12B). | STEP 6. INJECT CANA AT INJECTION SITE BY APPLYING EVEN PRESSURE TO THE INJECTOR, NOT A JABBING MOTION (FIGURE 7-14 OR 7-15). HOLD IN PLACE FOR 10 SECONDS. |
| STEP 7. CLEAR HARD OBJECTS FROM INJEC-TION SITE. | STEP 7. CLEAR HARD OBJECTS FROM INJEC-TION SITE. | STEP 7. BEND NEEDLE OF INJECTOR BY PRESSING ON A HARD SURFACE TO FORM A HOOK. |
| STEP 8. INJECT ATRO-PINE AT INJECTION SITE BY APPLYING EVEN PRESSURE TO THE IN-JECTOR, NOT A JABBING MOTION (FIGURE 7-10 OR 7-11). HOLD IN PLACE FOR 10 SECONDS. | STEP 8. INJECT ATNAA AT INJECTION SITE BY APPLYING EVEN PRES-SURE TO THE INJECTOR, NOT A JABBING MOTION (FIGURE 7-14 OR 7-15). HOLD IN PLACE FOR 10 SECONDS. | STEP 8. ATTACH USED INJECTOR TO BLOUSE POCKET FLAP OF BDO/JSLIST (FIGURE 7-16). |
| STEP 9. HOLD USED INJECTOR BETWEEN LITTLE FINGER AND RING FINGER OF NONDOM-INANT HAND (FIGURE 7-5A). | STEP 9. BEND NEEDLE OF INJECTOR BY PRESSING ON A HARD SURFACE TO FORM A HOOK. | STEP 9. MASSAGE INJECTION SITE, MISSION PERMITTING. |
| STEP 10. PULL LARGE INJECTOR (2 PAM CI) FROM CLIP (FIGURE 7-5C). DROP CLIP TO GROUND. | STEP 10. ATTACH ALL USED INJECTORS TO BLOUSE POCKET FLAP OF BDO/JSLIST (FIGURE 7-16). | |

*Table 7-2. Buddy Aid/Combat Lifesaver Aid for Nerve Agent Casualty (Continued).*

| MARK I* | ATNAA* | CANA** |
|---|---|---|
| STEP 11. INJECT 2 PAM CI AT INJECTION SITE BY APPLYING EVEN PRESSURE TO THE INJECTOR, NOT A JABBING MOTION (FIGURE 7-10 OR 7-11). HOLD IN PLACE FOR 10 SECONDS. | STEP 11. MASSAGE INJECTION SITE, MISSION PERMITTING. | |
| STEP 12. REPEAT STEPS ABOVE FOR REMAINING MARK Is. | | |
| STEP 13. BEND THE NEEDLES OF ALL USED INJECTORS BY PRESSING ON A HARD SURFACE TO FORM A HOOK. | | |
| STEP 14. ATTACH ALL USED INJECTORS TO BLOUSE POCKET FLAP OF BDO/JSLIST (FIGURE 7-13). | | |
| STEP 15. MASSAGE INJECTION SITE, MISSION PERMITTING. | | |

* USE STEPS LISTED FOR TYPE OF ANTIDOTE DEVICE ISSUED.

** CANA IS USED IN BUDDY AID/CLS AID ONLY. DO NOT USE IN SELF-AID.

**NOTE**

If the casualty is thinly built, inject the antidote into the buttock. Only inject the antidote into the upper outer portion of the casualty's buttock (Figure 7-11). This avoids hitting the nerve that crosses the buttocks (Figure 7-4). Hitting this nerve can cause paralysis.

*Figure 7-10. Injecting the casualty's thigh (Mark I or CANA).*

*Figure 7-11. Injecting the casualty's buttocks (Mark I or CANA).*

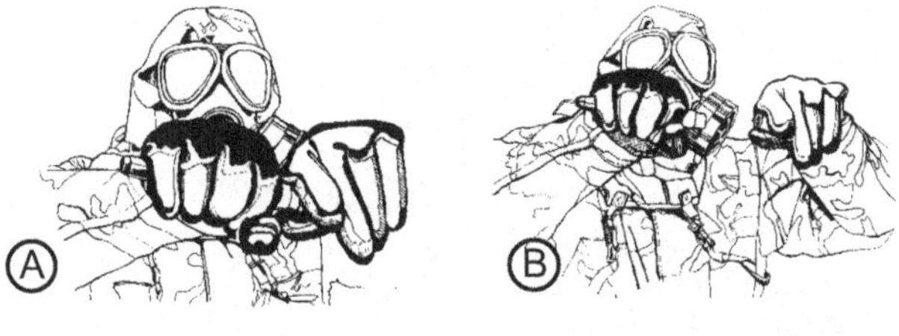


*Figure 7-12. Preparing CANA or ATNAA for injection.*

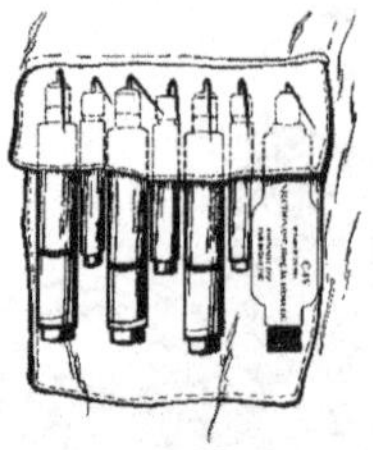

*Figure 7-13. Three sets of used MARK I autoinjectors and one CANA autoinjector attached to pocket flap.*

*d. Self-Administer Antidote Treatment Nerve Agent Autoinjector.* If you experience any or all of the nerve agent **MILD** symptoms (paragraph 7-7*b*), you must **IMMEDIATELY** self-administer one ATNAA following the procedure given Table 7-1.

**NOTE**

If you are thinly-built, inject yourself into the upper outer quarter (quadrant) of the buttock (Figure 7-15). There is a nerve that crosses the buttocks; hitting this nerve can cause paralysis. Therefore, you must only inject into the upper outer quarter (quadrant) of the buttocks.

*Figure 7-14. Self-administration of ATNAA (thigh).*

*Figure 7-15. Self-administration of ATNAA (buttock).*

NOTE

If you continue to have symptoms of nerve agent poisoning, seek someone else (a buddy) to check your symptoms and administer your remaining sets of injections, if required.

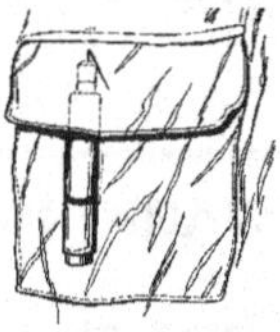

*Figure 7-16. Used ATNAA attached to clothing.*

*e. Buddy Assistance*. Service members may seek assistance after self-aid (self-administering one ATNAA) or may become incapacitated after self-aid. A buddy must evaluate the individual to determine if additional antidotes are required to counter the effects of the nerve agent. Also, service members may experience **SEVERE** symptoms of nerve agent poisoning (paragraph 7-7*b*); they will not be able to treat themselves. In either case, other service members must perform buddy aid as quickly as possible. Before initiating buddy aid, determine if one ATNAA has already been used so that no more than three ATNAA are administered. Buddy aid also includes administering the CANA with the third ATNAA to prevent convulsions. Follow the procedures indicated in Table 7-2.

**WARNING**

**Squat, DO NOT kneel, when masking the casualty or administering the nerve agent antidotes to the casualty. Kneeling may force any chemical agent on your overgarment into or through your protective clothing.**

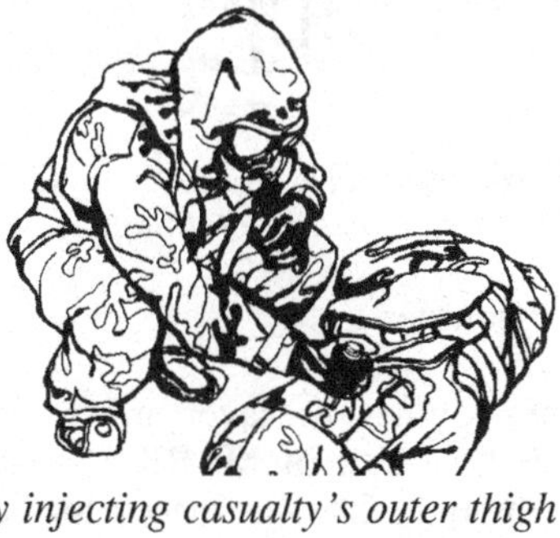

*Figure 7-17. Buddy injecting casualty's outer thigh (ATNAA or CANA).*

NOTE

If the casualty is thinly built, inject the antidote into the buttocks (Figure 7-18). Only inject the antidote into the upper outer portion of the casualty's buttocks. This avoids hitting the nerve that crosses the buttocks (Figure 7-4). Hitting this nerve can cause paralysis.

**WARNING**

**DO NOT inject into areas close to the hip, knee, or thighbone.**

*Figure 7-18. Buddy injecting casualty's buttocks (ATNAA or CANA).*

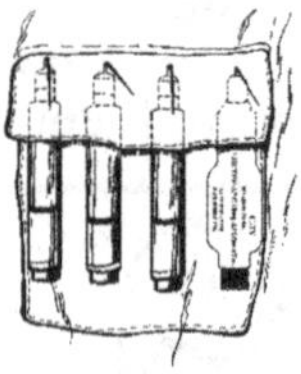

*Figure 7-19. Three used ATNAAs and one CANA autoinjector attached to clothing.*

*f. Combat Lifesaver.*

(1) The combat lifesaver must check to verify if the individual has received three sets of MARK I or ATNAAs. If not, the combat lifesaver performs first aid as described for buddy aid above. If the

individual has received the initial three sets of MARK I, then the combat lifesaver may administer additional atropine injections at approximately 15 minute intervals until atropinization is achieved (that is a heart rate above 90 beats per minute, reduced bronchial secretions, and reduced salivations). Administer additional atropine at intervals of 30 minutes to 4 hours to maintain atropinization or until the casualty is placed under the care of medical personnel. Check the heart rate by lifting the casualty's mask hood and feeling for a pulse at the carotid artery. Request medical assistance as soon as the tactical situation permits.

(2) The CLS should administer additional CANA to casualties suffering convulsions. Administer a second, and if needed, a third CANA at 5 to 10 minute intervals for a maximum of three injections (30 milligrams diazepam). Follow the steps and procedures described in buddy aid for administering the CANA. **DO NOT** give more than two additional injections for a total of three (one self-aid plus two by the CLS).

## 7-9. Blister Agents

Blister agents (vesicants) include mustard (H and HD), nitrogen mustards (HN), lewisite (L), and other arsenicals, mixtures of mustards and arsenicals, and phosgene oxime (CX). Blister agents may act on the eyes, mucous membranes, lungs, and skin. They burn and blister the skin or any other body parts they contact. Even relatively low doses may cause serious injury. Blister agents damage the respiratory tract (nose, sinuses, and windpipe) when inhaled and cause vomiting and diarrhea when absorbed. Lewisite and CX cause immediate pain on contact. However, mustard agents are deceptive as there is little or no pain at the time of exposure. Thus, in some cases, signs of injury may not appear for several hours after exposure.

*a. Protective Measures.* Your protective mask with hood and protective overgarment provide protection against blister agents. If it is known or suspected that blister agents are being used, **STOP BREATHING**, put on your mask and your protective overgarment.

**CAUTION**

Large drops of liquid vesicants on the protective overgarment ensemble may penetrate it if allowed to stand for an extended period. Remove large drops as soon as possible.

*b. Signs and Symptoms of Blister Agent Poisoning.*

(1) Immediate and intense pain upon contact with L, LH (lewisite and mustard) mixture, and CX. No initial pain upon contact with mustard.

(2) Inflammation and blisters (burns) resulting in tissue destruction. The severity of a chemical burn is directly related to the concentration of the agent and the duration of contact with the skin. The longer the agent is in contact with the tissue, the more serious the injury will be.

(3) Vomiting and diarrhea. Exposure to high concentrations of vesicants may cause vomiting or diarrhea.

(4) Death. The blister agent vapors absorbed during ordinary field exposure will probably not cause enough internal body (systemic) damage to result in death. However, death may occur from prolonged exposure to high concentrations of vapor or from extensive liquid contamination over wide areas of the skin, particularly *when decontamination is neglected or delayed.*

*c. First Aid Measures.*

(1) Use your M291 Skin Decontaminating Kit to decontaminate your skin and use water to flush contaminated eyes. Decontamination of vesicants must be done immediately (within 1 minute is best).

(2) If blisters form, cover them loosely with a field dressing and secure the dressing.

**CAUTION**

Blisters are actually burns. **DO NOT** attempt to decontaminate the skin where blisters have formed, as the agent has already been absorbed.

(3) If you receive blisters over a wide area of the body, you are considered seriously burned. Seek medical assistance immediately.

(4) If vomiting occurs, the mask should be lifted momentarily and drained—while the eyes are closed and the breath is held—and replaced, cleared, and sealed.

(5) Remember, if vomiting or diarrhea occurs after having been exposed to blister agents, seek medical assistance immediately.

## 7-10. Choking Agents (Lung-Damaging Agents)

Chemical agents that attack lung tissue, primarily causing fluid buildup (pulmonary edema), are classified as choking agents (lung-damaging agents). This group includes phosgene (CG), diphosgene (DP), chlorine (Cl), and chloropicrin (PS). Of these four agents, CG is the most dangerous and is more likely to be employed by the enemy in future conflict.

*a. Protective Measures.* Your protective mask gives adequate protection against choking agents.

*b. Signs and Symptoms.* During and immediately after exposure to choking agents (depending on agent concentration and length of exposure), you may experience some or all of the following signs and symptoms:

- Tears (lacrimation).
- Coughing.
- Choking.
- Tightness of chest.
- Nausea and vomiting.
- Headaches.

*c. Self-Aid.*

(1) The protective mask should be put on immediately when any of the conditions described in *b* above exist. Another indication of a CG attack is an odor like newly mown hay; however, **DO NOT** rely upon odor as indication of a chemical attack.

(2) If some CG is inhaled, normal combat duties should be continued unless there is difficulty in breathing, nausea, vomiting, or more than the usual shortness of breath during exertion. If any of the above symptoms occur and the mission permits, remain at quiet rest until medical evacuation is accomplished.

*d. Death.* With ordinary field exposure to choking agents, death will probably not occur. *However, prolonged exposure to high concentrations of the vapor and neglect or delay in masking can be fatal.*

## 7-11. Cyanogen (Blood) Agents

Cyanogen agents interfere with proper oxygen utilization in the body. Hydrogen cyanide (AC) and cyanogen chloride (CK) are the primary agents in this group.

*a. Protective Measures.* Your protective mask with a fresh filter gives adequate protection against field concentrations of cyanogen agent vapor. The protective overgarments, as well as the mask, are needed when exposed to liquid AC.

*b. Signs and Symptoms.* During and immediately after exposure to cyanogen agents (depending on agent concentration and length of exposure), you may experience some or all of the following signs and symptoms:

- Tearing (lacrimation).
- Eye, nose, and throat irritation.
- Sudden stimulation of breathing (unable to hold breath).
- Nausea.
- Coughing.
- Tightness of chest.
- Headache.
- Light-headedness (dizziness).
- Unconsciousness.

*c. First Aid.*

(1) *Hydrogen cyanide.* During any chemical attack, if you get a sudden stimulation of breath or detect an odor like bitter almonds, **PUT ON YOUR MASK IMMEDIATELY**. Speed is absolutely essential since this agent acts so rapidly that within a few seconds its effects will make it impossible for service members to put on their mask by themselves. Stop breathing until the mask is on, if at all possible. This may be very difficult since the agent strongly stimulates respiration.

(2) *Cyanogen chloride.* Put your mask on immediately if you experience any irritation of the eyes, nose, or throat. Service members

who are unable to mask should be masked by the nearest service member (buddy).

*d. Medical Assistance.* If you suspect that you have been exposed to blood agents, seek medical assistance immediately.

## 7-12. Incapacitating Agents

An incapacitating agent is a chemical agent which produces temporary, disabling conditions which persist for hours to days after exposure. Unlike riot control agents, which usually are momentary or fleeting in action, incapacitating agents have a persistent effect. It is likely that smoke-producing munitions or aerosols will disseminate such agents, thus making breathing their means of entry into the body. The protective mask is, therefore, essential.

*a.* There are no specific first aid measures to relieve the symptoms of incapacitating agents. Supportive first aid and physical restraint may be indicated. If the casualty is stuporous or comatose, be sure that respiration is unobstructed; then turn him on his side in case vomiting should occur. Complete cleansing of the skin with soap and water should be done as soon as possible; or, the M291 Skin Decontaminating Kit can be used if washing is impossible. Remove weapons and other potentially harmful items from service members who are suspected of having these symptoms. Harmful items include cigarettes, matches, medications, and small items that might be swallowed accidentally. Delirious (confused) persons have been known to attempt to eat items bearing only a superficial resemblance to food.

*b.* Incapacitating agents (anticholinergic drugs BZ type) may produce alarming dryness and coating of the lips and tongue; however, there is usually no danger of immediate dehydration. Fluids should be given sparingly, if at all, because of the danger of vomiting and because of the likelihood of temporary urinary retention due to paralysis of bladder muscles.

*c.* If the body temperature is elevated and mucous membranes are dry, immediate and vigorous cooling (as for heatstroke) is indicated. Methods that can be used to cool the skin are spraying with cool water and air circulation (fanning); applying alcohol soaked cloths and air circulation; and providing maximum exposure to air in a shaded area, along with maximum air circulation. Such cases are usually a result of anticholinergic poisoning. Rapid evacuation should be accomplished since medical treatment with the appropriate medication may be lifesaving.

**CAUTION**

**DO NOT** use **ice** for cooling the skin.

*d.* Reassurance and a firm, but friendly, attitude by individuals providing first aid will be beneficial if the casualty appears to comprehend what is being said. Conversation is a waste of time if the service member is incoherent or cannot understand what is being said. In such cases, the less said, the better it is—these casualties will benefit more from prompt and vigorous restraint and evacuation to an MTF.

## 7-13. Incendiaries

Incendiaries can be grouped as WP, thickened gasoline, metal, and oil and metal. You must learn to protect yourself against these incendiaries.

*a.* White phosphorus is used primarily as a smoke producer but can be used for its incendiary effect to ignite field expedients and combustible materials. The burns from WP are usually multiple, deep, and variable in size. When particles of WP get on the skin or clothing, they continue to burn until deprived of air. They also have a tendency to stick to a surface and must be brushed off or picked out.

(1) If burning particles of WP strike and stick to your clothing, quickly take off the contaminated clothing before the WP burns through to the skin.

(2) If burning WP strikes your skin, smother the flame with water, a wet cloth, or mud.

**NOTE**

Since WP is soluble in oil, **DO NOT** use grease, oily ointments, or eye ointments to smother the flame.

(3) Keep the WP particles covered with a wet material to exclude air until you can remove them or have them removed from your skin.

(4) Remove the WP particles from the skin by brushing them with a wet cloth and by picking them out with a knife, bayonet, stick, or other available object.

(5) Seek medical assistance when the mission permits.

*b.* Thickened fuel mixtures (napalm) have a tendency to cling to clothing and body surfaces, thereby producing prolonged exposure and severe burns. The first aid for these burns is the same as for other heat burns. The heat and irritating gases given off by these combustible mixtures may cause lung damage, which must be treated by medical personnel.

*c.* Metal incendiaries pose special problems. Thermite particles on the skin should be immediately cooled with water and then removed. The first aid for these burns is the same as for other heat burns. Particles of magnesium on the skin burn quickly and deeply. Like other metal incendiaries, they must be removed. Ordinarily, medical personnel should do the complete removal of these particles as soon as possible. Immediate medical treatment is required.

*d.* Oil and metal incendiaries have much the same effect on contact with the skin and clothing as those discussed (*b* and *c* above). First aid measures for burns are discussed in Chapter 3.

## 7-14. Biological Agents and First Aid

*a.* Biological attacks can result in combat ineffectiveness by introducing disease-causing organisms into a troop population.

*b.* Once a disease is identified, first aid or medical treatment is initiated, depending on the seriousness of the disease. First aid measures are concerned with observable symptoms of the disease such as diarrhea or vomiting.

## 7-15. Toxins

Toxins are alleged to have been used in past conflicts. Witnesses and victims have described the agent as toxic rain (or yellow rain) because it was reported to have been released from aircraft as a yellow powder or liquid that covered ground, structures, vegetation, and people.

*a.* *Signs and Symptoms.* The occurrence of the symptoms from toxins may appear in a period of a few minutes to several hours depending on the particular toxin, the service member's susceptibility, and the amount of toxin inhaled, ingested, or deposited on the skin. Symptoms from toxins usually involve the central nervous system but are often preceded by less prominent symptoms, such as nausea, vomiting, diarrhea, cramps, or stomach irritation and burning sensation. Typical neurological symptoms often develop rapidly in severe cases; for example, visual disturbances, inability to swallow, speech difficulty, lack of muscle coordination, and sensory abnormalities (numbness of mouth, throat, or extremities). Yellow rain (mycotoxins) also may have hemorrhagic symptoms, which could include any or all of the following:

- Dizziness.
- Severe itching or tingling of the skin.
- Formation of multiple, small, hard blisters.
- Coughing up blood.
- Shock (which could result in death).

*b.* *Self-Aid.* Upon recognition of an attack employing toxins, you must immediately take the following actions:

(1) Stop breathing, put on your protective mask with hood, and then resume breathing. Next, put on your protective clothing.

(2) Should severe itching of the face become unbearable, quickly—

- Loosen the cap on your canteen.
- Take and hold a deep breath and lift your mask.
- While holding your breath, close your eyes and flush your face with generous amounts of water.

**CAUTION**

**DO NOT** rub or scratch your eyes. Try not to let the water run onto your clothing or protective overgarment.

- Put your protective mask back on, seat it properly, clear it, and check it for a seal; then resume breathing.
- Decontaminate your skin by bathing with soap and water as soon as the mission permits.

- Change clothing and decontaminate your protective mask using soap and water. Replace the filters if directed.

(3) If vomiting occurs, the mask should be lifted momentarily and drained—while the eyes are closed and the breath is held—and replaced, cleared, and sealed.

*c.* *Medical Assistance.* If you suspect that you have been exposed to toxins, you should seek medical assistance immediately.

**7-16. Nuclear Detonation**

*a.* Three types of injuries may result from a nuclear detonation. These are thermal, blast, and radiation injuries. Many times the casualty will have a combination of these types of injuries. First aid for thermal and blast injuries is provided based on observable injuries, such as burns, hemorrhage, or fractures.

*b.* The signs and symptoms of radiation illness in the initial phase include the rapid onset of nausea, vomiting, and malaise (tiredness). The only first aid procedure for radiological casualties is decontamination.

CHAPTER 8

# FIRST AID FOR PSYCHOLOGICAL REACTIONS

## 8-1. General

Psychological first aid is as natural and reasonable as physical first aid and is just as familiar. When you were hurt as a child, the understanding attitude of your parents did as much as the psychological effect of a bandage. Later, your disappointment or grief was eased by supportive words from a friend. Certainly, taking a walk and talking things out with a friend are familiar ways of dealing with an emotional crisis. The same natural feelings that make us want to help a person who is injured make us want to give a helping hand to a buddy who is upset. *Psychological first aid* really means nothing more complicated than assisting people with emotional distress whether it results from physical injury, disease, or excessive stress. Emotional distress is not always as visible as a wound or a broken bone. However, overexcitement, severe fear, excessive worry, deep depression, misdirected irritability, and anger are signs that stress has reached the point of interfering with effective coping. The more noticeable the symptoms become, the more urgent the need for you to be of help and the more important it is for you to know *how* to help.

## 8-2. Importance of Psychological First Aid

You must know how to give psychological first aid to be able to help yourself, your buddies, and your unit in order to keep performing the mission. Psychological first aid measures are simple and easy to understand. Your decision of what to do depends upon your ability to observe the service member and understand his needs. Making the best use of resources requires ingenuity on your part. A stress reaction resulting in poor judgment can cause injury or even death to yourself or others on the battlefield. It can be even more dangerous if other persons are affected by the judgment of an emotionally upset service member. If it is detected early enough, the affected service member stands a good chance of remaining in his unit as an effective member. If it is not detected early and if the service member becomes more emotionally upset, he may become a threat to himself and to others.

## 8-3. Situations Requiring Psychological First Aid

- Psychological first aid (buddy aid) is most needed at the first sign that a service member cannot perform the mission because of emotional distress. Stress is inevitable in combat, in hostage and terrorist situations, and in civilian disasters such as floods, hurricanes, or industrial accidents. Most emotional reactions to such situations are temporary, and the service member can still carry on with encouragement. Painful or disruptive symptoms may last for minutes, hours, or days. However, if the stress symptoms are seriously disabling, they may be psychologically contagious and endanger not only the emotionally upset service member but also the entire unit.

- Sometimes people continue to function well during a disastrous event, but suffer from emotional scars which impair their job performance or quality of life at a later time. Painful memories and dreams may recur for months and years and still be considered a normal reaction. However, if the memories are so painful that the person must avoid all situations which arouse them, becomes socially withdrawn, or shows symptoms of anxiety, depression, or substance abuse, he needs treatment. Experience with police, firemen, emergency medical technicians, and others who deal with disasters has proved that the routine application of psychological first aid to all the participants, including those who have functioned well, greatly reduces the likelihood of future serious post-traumatic stress disorders (PTSDs).

## 8-4. Interrelationship of Psychological and Physical First Aid

Psychological first aid should go hand in hand with physical first aid. The discovery of a physical injury or cause for an inability to function does not rule out the possibility of a psychological injury (or vice versa). The person suffering from pain, shock, fear of serious injury, or fear of death does not respond well to joking, indifference, or fearful-tearful attention. Fear and anxiety may take as high a toll of the service member's strength as does the loss of blood.

## 8-5. Goals of Psychological First Aid

The goals of psychological first aid are to—

- Be supportive; assist the service member in dealing with his stress reaction.

- Prevent, and if necessary control, behavior harmful to himself and to others.

- Return the service member to duty as soon as possible after dealing with the stress reaction.

## 8-6. Respect for Others' Feelings

*a.* Accept the service member you are trying to help without censorship or ridicule. Respect his right to his own feelings. Even though your feelings, beliefs, and behavior are different, DO NOT blame or make light of him for the way he feels or acts. Your purpose is to help him in this tough situation, not to be his critic. A person DOES NOT WANT to be upset and worried. When he seeks help, he needs and expects consideration of his fears, not abrupt dismissal or ridicule.

*b.* Realize that people are the products of a wide variety of factors. All people DO NOT react the same way to the same situations. Each individual has complex needs and motivations, both conscious and unconscious, that are uniquely his own. Often the one thing that finally causes the person to become overloaded by a stressful situation is not the stressor itself, but some other problem.

## 8-7. Emotional and Physical Disability

*a.* Accept emotional disability as being just as real as physical disability. If a service member's ankle is seriously sprained in a fall, no one expects him to run right away. A service member's emotions may be temporarily strained by the overwhelming stress of battle or other traumatic incident. DO NOT demand that he pull himself together immediately and carry on without a break. Some individuals can pull themselves together immediately, but others cannot. The service member whose emotional stability has been disrupted has a disability just as real as the service member who has sprained his ankle. There is an unfortunate tendency in many people to regard as real only what they can see, such as a wound or bleeding. Some people tend to assume that damage involving a person's mind and emotions is just imagined, that he is not really sick or injured, and that he could overcome his trouble by using his will power.

*b.* The terms *it's all in your head, snap out of it,* and *get control of yourself* are often used by people who believe they are being helpful. Actually, these terms are expressions of hostility because they show lack of understanding. They only emphasize weakness and inadequacy. Such terms are of no use in psychological first aid.

*c.* Every physically injured person has some emotional reaction to the fact that he is injured.

(1) It is normal for an injured person to feel upset. The more severe the injury, the more insecure and fearful he becomes, especially

if the injury is to a body part which is highly valued. For example, an injury to the eyes or the genitals, even though relatively minor, is likely to be extremely upsetting. An injury to some other part of the body may be especially disturbing to an individual for his own particular reason. For example, an injury of the hand may be a terrifying blow to a surgeon or an injury to the eye of a pilot.

(2) An injured service member always feels less secure, more anxious, and more afraid not only because of what has happened to him but because of what he imagines may happen as a result of his injury. This fear and insecurity may cause him to be irritable, uncooperative, or unreasonable. As you help him, always keep in mind that such behavior has little or nothing to do with you personally. He needs your patience, reassurance, encouragement, and support.

## 8-8. Combat and Other Operational Stress Reactions

*Stress reaction* is a temporary emotional disorder or inability to function, experienced by a previously normal service member as a reaction to the overwhelming or cumulative stress of combat. Stress reaction gets better with reassurance, rest, physical replenishment, and activities that restore confidence. All service members are likely to feel stress reaction under conditions of intense and/or prolonged stress. They may even become stress reaction casualties, unable to perform their mission for hours or days. Other combat and operational stress reactions (COSRs) may result in negative behavior, but are not termed *stress reaction,* as they need more intensive treatment. These negative COSRs may result in misconduct stress behaviors such as drug and alcohol abuse, criminal acts, looting, desertion, and self-inflicted wounds. These harmful COSRs can often be prevented by good psychological first aid. Service members who commit misconduct stress behaviors may require disciplinary action rather than medical treatment.

## 8-9. Reactions to Stress

Most service members react to stressful incidents after the situation has passed. All service members feel some fear. This fear may be greater than they have experienced at any other time, or they may be more aware of their fear. In such a situation, they should not be surprised if they feel shaky or become sweaty, nauseated, or confused. These reactions are normal and are not a cause for concern. However, some reactions, either short- or long-term, will cause problems if left unchecked. See paragraph 8-13 for more information.

*a. Emotional Reactions.*

(1) The most obvious combat stress reaction (CSR) is inefficient performance. This can be demonstrated by—

- Slow thinking (or reaction time).
- Difficulty recognizing priorities and seeing what needs to be done.
- Difficulty getting started.
- Indecisiveness and having trouble focusing attention.
- Tendency to do familiar tasks and be preoccupied with familiar details. (This can reach the point where the person is very passive, such as just sitting or wandering about not knowing what to do.)

(2) A less common reaction may be uncontrolled emotional outbursts; this can be demonstrated by crying, screaming, or laughing. Some service members will react in the opposite way. They will be very withdrawn and silent and try to isolate themselves from everyone. These service members should be encouraged to remain with their assigned unit. Uncontrolled reactions may appear by themselves or in any combination (the person may be crying uncontrollably one minute and then laughing the next). In this state, the person is restless and cannot keep still. He may run about, apparently without purpose. Inside, he feels a great rage or fear and his physical acts may show this. In his anger he may indiscriminately strike out at others.

*b. Loss of Adaptability.*

(1) In a desperate attempt to get away from the danger, which has overwhelmed him, a service member may panic and become confused. His mental ability may be so impaired he cannot think clearly or even follow simple commands. His judgment may be faulty and he may not be aware of his actions, such as standing up in his fighting position during an attack.

(2) In other cases, overwhelming stress may produce symptoms that are often associated with head injuries. For example, the service member may appear dazed or be found wandering around aimlessly. He may appear confused and disoriented and may seem to have a complete or partial loss of memory. In such cases, especially when no eyewitnesses can provide evidence that the service member has NOT suffered a head injury, it is necessary for him to be rapidly medically evacuated. **DO NOT** allow the

service member to expose himself to further personal danger until the cause of the problem has been determined.

*c. Sleep Disturbance and Repetition of Dreams.* A person who has been overwhelmed by stress often has difficulty sleeping. The service member may experience nightmares related to the stressors. Remember that nightmares, in themselves, are not considered abnormal when they occur soon after a period of intensive stress. As time passes, the nightmares usually become less frequent and less intense. In extreme cases, a service member, even when awake, may think repeatedly of the incident, feel as though it is happening again, and act out parts of his stress over and over again. For some persons, this repetitious reexperiencing of the stressful event may be necessary for eventual recovery; therefore, it should not be discouraged or viewed as abnormal. For the person reexperiencing the event, such reaction may be disruptive. The service member needs to be encouraged to *ventilate* about the incident. Ventilation is a technique where the service member is given the opportunity to talk extensively, often repetitiously about the experience.

## 8-10. Severe Stress or Stress Reaction

You do not need specialized training to recognize severe stress or stress reaction that will cause problems for the service member, the unit, or the mission. Reactions that are less severe, however, are more difficult to detect. To determine whether a person needs help, you must observe him to see whether he is doing something meaningful, performing his duties, taking care of himself, behaving in an unusual fashion, or acting out of character.

## 8-11. Application of Psychological First Aid

The emotionally disturbed service member has built a barrier against fear. He does this for his own protection, although he is probably not aware that he is doing it. If he finds that he does not have to be afraid and that there are normal, understandable things about him, he will feel safer in dropping this barrier. Persistent efforts to make him realize that you want to understand him will be reassuring, especially if you remain calm. Nothing can cause an emotionally disturbed person to become even more fearful than feeling that others are afraid of him. Try to remain calm. Familiar things, such as a cup of coffee, the use of his name, attention to a minor wound, being given a simple job to do, or the sight of familiar people and activities, will add to his ability to overcome his fear. He may not respond well if you get excited, angry, or abrupt.

*a.* *Ventilation.* After the service member becomes calmer, he is likely to have dreams about the stressful event. He also may think about it when he is awake or even repeat his personal reaction to the event. One benefit of this natural pattern is that it helps him master the stress by going over it just as one masters the initial fear of parachuting from an aircraft by doing it over and over again. Eventually, it is difficult to remember how frightening the event was initially. In giving first aid to the emotionally disturbed service member, you should let him follow this natural pattern. Encourage him to talk. Be a good listener. Let him tell, in his own words, what actually happened. If home front problems or worries have contributed to the stress, it will help him to talk about them. Your patient listening will prove to him that you are interested in him, and by describing his personal problem, he can work at mastering his fear. If he becomes overwhelmed in the telling, suggest a cup of coffee or a break. Whatever you do, assure him that you will listen again as soon as he is ready. Do try to help put the service member's perception of what happened back into realistic perspective; but DO NOT argue about it.

*b.* *Activity.*

(1) A person who is emotionally disturbed as the result of a combat action is a casualty of anxiety and fear. He is disabled because he has become temporarily overwhelmed by his anxiety. A good way to control fear is through activity. Almost all service members, for example, experience a considerable sense of anxiety and fear while they are poised, awaiting the opening of a big offensive; but this is normally relieved, and they actually feel better once they begin to move into action. They take pride in effective performance and pleasure in knowing that they are good service members, perhaps being completely unaware that overcoming their initial fear was their first major accomplishment.

(2) Useful activity is very beneficial to the emotionally disturbed service member who is not physically incapacitated. After you help a service member get over his initial fear, help him to regain some self-confidence. Make him realize his job is continuing by finding him something useful to do. Encourage him to be active. Get him to help load trucks, clean up debris, or dig fighting positions. If possible, get him back to his usual duty. Seek out his strong points and help him apply them. Avoid having him just sit around. You may have to provide direction by telling him what to do and where to do it. The instructions should be clear and simple and should be repeated. A person who has panicked is likely to argue. Respect his feelings, but point out more immediate, obtainable, and demanding needs. Channel his excessive energy and, above all, DO NOT argue. If you cannot get him interested in doing more profitable work, it may be necessary to enlist aid in controlling his overactivity before it spreads to the group and

results in more panic. Prevent the spread of such infectious feelings by restraining and segregating if necessary.

(3) Involvement in activity helps a service member in three ways; he—

- Forgets himself.
- Has an outlet for his excessive tensions.
- Proves to himself he can do something useful.

*c.* *Rest.* There are times, particularly in combat, when physical exhaustion is a principal cause for emotional reactions. A unit sleep plan should be established and implemented. When possible, service members should be given a safe and relatively comfortable area in which to sleep. Examples would be an area away from heavy traffic, noise, and congestion or a place that is clean and dry and protected from environmental conditions. The more uninterrupted sleep a service member gets the better he will be able to function in the tactical environment.

*d.* *Hygiene.* Field hygiene is an important ingredient in a service member's morale. A service member who is dirty and unkempt will not function as well as a service member who has had the opportunity to bathe and put on clean, dry clothing. During combat, unit leaders should stress the importance of personal hygiene. Good personal hygiene not only improves morale, it also is a preventive measure against disease and nonbattle injury (DNBI).

*e.* *Group Activity.* You have probably already noticed that a person works, faces danger, and handles serious problems better if he is a member of a closely-knit group. Each service member in the team supports the other team members. Esprit de corps is built because the service members have the same interests, goals, and mission, and as a result they are more productive; furthermore, they are less worried because everyone is involved. It is this spirit that takes a strategic hill in battle. It is so powerful that it is one of the most effective tools you have in your *psychological first aid bag*. Getting the service member back into the team or squad activities will reestablish his sense of belonging and security and will go far toward making him a useful member of the unit.

## 8-12. Reactions and Limitations

Up to this point the discussion has been primarily about the feelings of the emotionally distressed service member. What about your feelings toward him? Whatever the situation, you will have emotional reactions (conscious or unconscious) toward this service member. Your reactions can either help or hinder your ability to help him. When you are tired or worried, you may very easily become impatient with him if he is unusually slow or exaggerates. You may even feel resentful toward him. At times when many physically wounded lie about you, it will be especially natural for you to resent disabilities that you cannot see. Physical wounds can be seen and easily accepted. Emotional reactions are more difficult to accept as injuries. On the other hand, will you tend to be overly sympathetic? Excessive sympathy for an incapacitated person can be as harmful as negative feelings in your relationship with him. He needs strong help, but not your sorrow. To overwhelm him with pity will make him feel even more inadequate. You must expect your buddy to recover, to be able to return to duty, and to become a useful service member again. This expectation should be displayed in your behavior and attitude as well as in what you say. If he can see your calmness, confidence, and competence, he will be reassured and will feel a sense of greater security.

## 8-13. Stress Reactions

See Tables 8-1, 8-2, and 8-3 for more information.

*Table 8-1. Mild Stress Reaction*

| PHYSICAL SIGNS* | EMOTIONAL SIGNS* |
|---|---|
| 1. TREMBLING, TEARFUL<br>2. JUMPINESS, NERVOUSNESS<br>3. COLD SWEAT, DRY MOUTH<br>4. POUNDING HEART, DIZZINESS<br>5. INSOMNIA, NIGHTMARES<br>6. NAUSEA, VOMITING, DIARRHEA<br>7. FATIGUE<br>8. THOUSAND-YARD STARE<br>9. DIFFICULTY THINKING, SPEAKING, AND COMMUNICATING | 1. ANXIETY, INDECISIVENESS<br>2. IRRITABLE, COMPLAINING<br>3. FORGETFUL, UNABLE TO CONCENTRATE<br>4. EASILY STARTLED BY NOISE, MOVEMENT<br>5. GRIEF, TEARFUL<br>6. ANGER, BEGINNING TO LOSE CONFIDENCE IN SELF AND UNIT |

| SELF- AND BUDDY AID |
|---|
| 1. CONTINUE MISSION PERFORMANCE, FOCUS ON IMMEDIATE MISSION.<br>2. EXPECT SERVICE MEMBER TO PERFORM ASSIGNED DUTIES.<br>3. REMAIN CALM AT ALL TIMES; BE DIRECTIVE AND IN CONTROL.<br>4. LET SERVICE MEMBER KNOW HIS REACTION IS NORMAL, AND THAT THERE IS NOTHING SERIOUSLY WRONG WITH HIM.<br>5. KEEP SERVICE MEMBER INFORMED OF THE SITUATION, OBJECTIVES, EXPECTATIONS, AND SUPPORT. CONTROL RUMORS.<br>6. BUILD SERVICE MEMBER'S CONFIDENCE, TALK ABOUT SUCCEEDING.<br>7. KEEP SERVICE MEMBER PRODUCTIVE (WHEN NOT RESTING) THROUGH RECREATIONAL ACTIVITIES, EQUIPMENT MAINTENANCE. |

8. ENSURE SERVICE MEMBER MAINTAINS GOOD PERSONAL HYGIENE.
9. ENSURE SERVICE MEMBER EATS, DRINKS, AND SLEEPS AS SOON AS POSSIBLE.
10. LET SERVICE MEMBER TALK ABOUT HIS FEELINGS. DO NOT "PUT DOWN" HIS FEELINGS OF GRIEF OR WORRY. GIVE PRACTICAL ADVICE AND PUT EMOTIONS INTO PERSPECTIVE.

* MOST OR ALL OF THESE SIGNS ARE PRESENT IN MILD STRESS REACTION. THEY CAN BE PRESENT IN ANY NORMAL SERVICE MEMBER IN COMBAT YET HE CAN STILL DO HIS JOB.

*Table 8-2. More Serious Stress Reaction*

| PHYSICAL SIGNS* | EMOTIONAL SIGNS* |
|---|---|
| 1. CONSTANTLY MOVES AROUND | 1. RAPID AND/OR INAPPROPRIATE TALKING |
| 2. FLINCHING OR DUCKING AT SUDDEN SOUNDS | 2. ARGUMENTATIVE, RECKLESS MOVEMENTS/ACTIONS |
| 3. SHAKING, TREMBLING (WHOLE BODY OR ARMS) | 3. INATTENTIVE TO PERSONAL HYGIENE |
| 4. CANNOT USE PART OF BODY, NO PHYSICAL REASON (HAND, ARM, LEGS) | 4. INDIFFERENT TO DANGER |
| 5. CANNOT SEE, HEAR, OR FEEL (PARTIAL OR COMPLETE LOSS) | 5. MEMORY LOSS |
| 6. PHYSICAL EXHAUSTION, CRYING | 6. SEVERE STUTTERING, MUMBLING, OR CANNOT SPEAK AT ALL |
| 7. FREEZING UNDER FIRE, OR TOTAL IMMOBILITY | 7. INSOMNIA, NIGHTMARES |
| 8. VACANT STARES, STAGGERS, SWAYS WHEN STANDS | 8. SEEING OR HEARING THINGS THAT DO NOT EXIST |
| 9. PANIC RUNNING UNDER FIRE | 9. RAPID EMOTIONAL SHIFTS |
| | 10. SOCIAL WITHDRAWAL |
| | 11. APATHETIC |
| | 12. HYSTERICAL OUTBURSTS |
| | 13. FRANTIC OR STRANGE BEHAVIOR |

**TREATMENT PROCEDURES****

1. IF A SERVICE MEMBER'S BEHAVIOR ENDANGERS THE MISSION, SELF, OR OTHERS, DO WHATEVER IS NECESSARY TO CONTROL HIM.
2. IF THE SERVICE MEMBER IS UPSET, CALMLY TALK HIM INTO COOPERATING.
3. IF CONCERNED ABOUT THE SERVICE MEMBER'S RELIABILITY:
   - UNLOAD HIS WEAPON.
   - TAKE WEAPON IF SERIOUSLY CONCERNED.
   - PHYSICALLY RESTRAIN HIM ONLY WHEN NECESSARY FOR SAFETY OR TRANSPORTATION.
4. REASSURE EVERYONE THAT THE SIGNS ARE PROBABLY JUST STRESS REACTION AND WILL QUICKLY IMPROVE.
5. IF STRESS REACTION SIGNS CONTINUE:
   - GET THE SERVICE MEMBER TO A SAFER PLACE.
   - DO NOT LEAVE THE SERVICE MEMBER ALONE, KEEP SOMEONE HE KNOWS WITH HIM.
   - NOTIFY SENIOR NONCOMMISSIONED OFFICER (NCO) OR OFFICER.
   - HAVE THE SERVICE MEMBER EXAMINED BY MEDICAL PERSONNEL.

*Table 8-2. More Serious Stress Reaction (Continued)*

| TREATMENT PROCEDURES** |
|---|
| 6. GIVE THE SERVICE MEMBER EASY TASKS TO DO WHEN NOT SLEEPING, EATING, OR RESTING.<br>7. ASSURE THE SERVICE MEMBER HE WILL RETURN TO FULL DUTY IN 24 HOURS; AND, RETURN HIM TO NORMAL DUTIES AS SOON AS HE IS READY. |
| * THESE SIGNS ARE PRESENT IN ADDITION TO THE SIGNS OF MILD STRESS REACTION.<br>** DO THESE PROCEDURES IN ADDITION TO THE SELF- AND BUDDY AID CARE. |

*Table 8-3. Preventive Measures to Combat Stress Reaction*

1. WELCOME NEW MEMBERS INTO YOUR TEAM, GET TO KNOW THEM QUICKLY. IF YOU ARE NEW, BE ACTIVE IN MAKING FRIENDS.
2. BE PHYSICALLY FIT (STRENGTH, ENDURANCE, AND AGILITY).
3. KNOW AND PRACTICE LIFESAVING SELF- AND BUDDY AID.
4. PRACTICE RAPID RELAXATION TECHNIQUES (FM 22-51).
5. HELP EACH OTHER OUT WHEN THINGS ARE TOUGH AT HOME OR IN THE UNIT.
6. KEEP INFORMED; ASK YOUR LEADER QUESTIONS, IGNORE RUMORS.
7. WORK TOGETHER TO GIVE EVERYONE FOOD, WATER, SHELTER, HYGIENE, AND SANITATION.
8. SLEEP WHEN MISSION AND SAFETY PERMIT; LET EVERYONE GET TIME TO SLEEP.
   - SLEEP ONLY IN SAFE PLACES AND BY STANDING OPERATING PROCEDURE (SOP).
   - IF POSSIBLE, SLEEP 6 TO 9 HOURS PER DAY.
   - TRY TO GET AT LEAST 4 HOURS SLEEP PER DAY.
   - GET GOOD SLEEP BEFORE GOING ON SUSTAINED OPERATIONS.
   - CATNAP WHEN YOU CAN, BUT ALLOW TIME TO WAKE UP FULLY.
   - CATCH UP ON SLEEP AFTER GOING WITHOUT.

APPENDIX A

# FIRST AID CASE AND KITS, DRESSINGS, AND BANDAGES

## A-1. First Aid Case with Field Dressings and Bandages

Every service member is issued a first aid case (Figure A-1A) with a field first aid dressing encased in a plastic wrapper (Figure A-1B). He carries it at all times for his use. The field first aid dressing is a standard sterile (germ-free) compress or pad with bandages attached (Figure A-1C). This dressing is used to cover the wound, to protect against further contamination, and to stop bleeding (pressure dressing). When a service member administers first aid to another person, he must remember to use the wounded person's dressing; he may need his own later. The service member must check his first aid case regularly and replace any used or missing dressing. The field first aid dressing may normally be obtained from his unit supply.

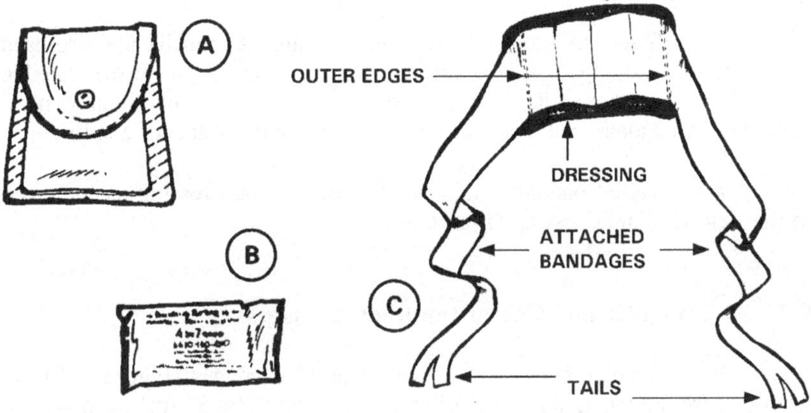


*Figure A-1. Field first aid case and dressing (Illustrated A—C).*

## A-2. General Purpose First Aid Kits

General-purpose first aid kits are listed in the common table of allowances (CTA) 8-100. The operators, crew, and passengers carry these kits on Army vehicles, aircraft, and boats for use. Individuals designated by unit TSOP to be responsible for these kits are required to check them regularly and replace all items used. The general-purpose kit and its contents can be obtained through the unit supply system.

**NOTE**

Periodically check the dressings (for holes or tears in the packaging) and the medicines (for expiration date) that are in the first aid kits. If necessary, replace defective or outdated items.

## A-3. Dressings

Dressings are sterile pads or compresses used to cover wounds. They usually are made of gauze or cotton wrapped in gauze (Figure A-1C). In addition to the standard field first aid dressing, other dressings such as sterile gauze compresses and small sterile compresses on adhesive strips may be available under CTA 8-100.

## A-4. Standard Bandages

*a.* Standard bandages are made of gauze or muslin and are used over a sterile dressing to secure the dressing in place, to close off its edge from dirt and germs, and to create pressure on the wound and control bleeding. A bandage can also support an injured part or secure a splint.

*b.* Tailed bandages may be attached to the dressing as indicated on the field first aid dressing (Figure A-1C).

## A-5. Triangular and Cravat (Swathe) Bandages

*a.* Triangular and cravat (or swathe) bandages (Figure A-2) are fashioned from a triangular piece of muslin (37 by 37 by 52 inches) provided in the general-purpose first aid kit. If it is folded into a strip, it is called a cravat. Two safety pins are packaged with each bandage. These bandages are valuable in an emergency since they are easily applied.

*b.* To improvise a triangular bandage, cut a square of available material, slightly larger than 3 feet by 3 feet, and *fold it diagonally*. If two bandages are needed, cut the material along the diagonal fold.

*c.* A cravat can be improvised from such common items as T-shirts, other shirts, bed linens, trouser legs, scarfs, or any other item made of pliable and durable material that can be folded, torn, or cut to the desired size.

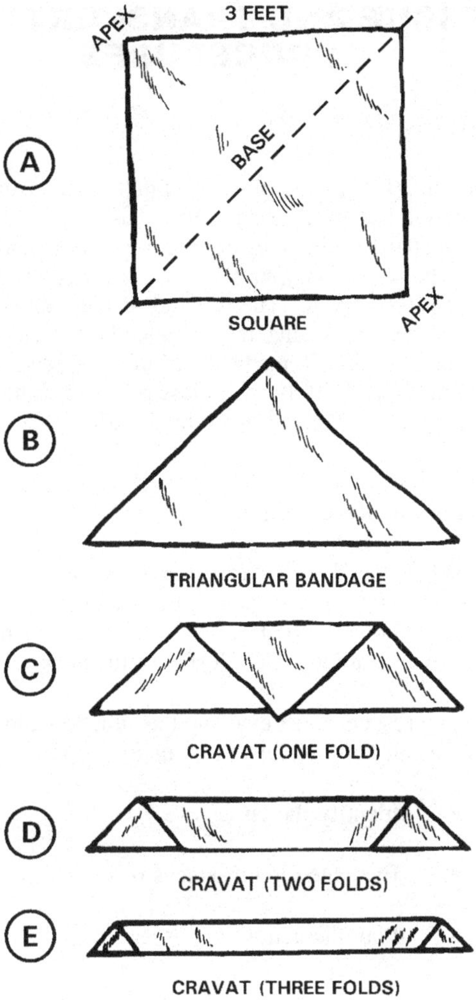


*Figure A-2. Triangular and cravat bandages (Illustrated A—E).*

APPENDIX B

# RESCUE AND TRANSPORTATION PROCEDURES

## B-1. General

A basic principle of first aid is to evaluate the casualty's injuries and administer first aid before moving him. However, adverse situations or conditions may jeopardize the lives of both the rescuer and the casualty if this is done. It may be necessary *first to rescue* the casualty before first aid can be effectively or safely given. The life and/or the well-being of the casualty will depend as much upon the manner in which he is *rescued and transported,* as it will upon the first aid and medical treatment he receives. Rescue actions must be done quickly and safely. Careless or rough handling of the casualty during rescue operations can aggravate his injuries.

## B-2. Principles of Rescue Operations

*a.* When faced with the necessity of rescuing a casualty who is threatened by hostile action, fire, water, or any other immediate hazard, DO NOT take action without first determining the extent of the hazard and your ability to handle the situation. DO NOT become a casualty.

*b.* The rescuer must evaluate the situation and analyze the factors involved. This evaluation involves three major steps:

- Identify the task.
- Evaluate circumstances of the rescue.
- Plan the action.

## B-3. Considerations

*a.* First determine if a rescue attempt is actually needed. It is a waste of time, equipment, and personnel to rescue someone not in need of rescuing. It is also a waste to look for someone who is not lost or needlessly risk the lives of the rescuer(s). In planning a rescue, attempt to obtain the following information:

- Who, what, where, when, why, and how the situation happened?

- How many casualties are involved and the nature of their injuries?

- What is the tactical situation?

- What are the terrain features and the location of the casualties?

- Will there be adequate assistance available to aid in the rescue/evacuation?

- Can first aid and/or medical treatment be provided at the scene; will the casualties require movement to a safer location?

- What specialized equipment will be required for the rescue operation?

- Is the rescue area contaminated? Will decontamination equipment and materiel be required for casualties, rescue personnel, and rescue equipment?

- How much time is available?

*b.* The time element can play a significant role in how the rescue is attempted. If the casualties are in imminent danger of losing their lives (such as near a burning vehicle or in a burning building) the time available will be relatively short and will sometimes cause a rescuer to compromise planning stages and/or the first aid which can be given. However, if the casualty is in a relatively secure area and his physical condition is strong, more deliberate planning can take place. A realistic estimate of time available must be made as quickly as possible to determine action time remaining. The key elements are the casualty's physical and mental condition, the tactical situation, and the environment.

## B-4. Plan of Action

*a.* The casualty's ability to endure is of primary importance in estimating the time available. Age, physical condition, and extent of wounds and/or injuries will differ from casualty to casualty. Therefore, to determine the time available, you will have to consider—

- Endurance time of the casualty.

- Extent of injuries.

- Type of situation.
- Personnel and/or equipment availability.
- Weather.
- Terrain (natural and man-made).
- Environment (contaminated or uncontaminated).

*b.* In respect to terrain, you must consider altitude and visibility. In some cases, the casualty may be of assistance because he knows more about the particular terrain or situation than you do. Maximum use of secure/reliable trails or roads is essential.

*c.* When taking weather into account, ensure that blankets and/or rain gear are available. Even a mild rain can complicate a normally simple rescue. In high altitudes and/or extreme cold and gusting winds, the time available is critically shortened. Be prepared to provide shelter and warmth for the casualty as well as the rescuers.

## B-5. Proper Handling of Casualties

*a.* You may have saved the casualty's life through the application of appropriate first aid measures. However, his life can be lost through rough handling or careless transportation procedures. Before you attempt to move the casualty—

- Evaluate the type and extent of his injuries.
- Ensure that dressings over wounds are adequately reinforced.
- Ensure that fractured bones are properly immobilized and supported to prevent them from cutting through muscle, blood vessels, and skin.

*b.* Based upon your evaluation of the type and extent of the casualty's injury and your knowledge of the various manual carries, you must select the best possible method of manual transportation. If the casualty is conscious, tell him how he is to be transported. This will help allay his fear of movement and gain his cooperation and confidence.

*c.* Buddy aid for chemical agent casualties includes those actions required to prevent an incapacitated casualty from receiving additional injury from the effects of chemical hazards. If a casualty is physically unable to decontaminate himself or administer the proper chemical agent antidote, the casualty's buddy assists him and assumes responsibility for his care. Buddy-aid includes—

- Administering the proper chemical agent antidote.
- Decontaminating the incapacitated casualty's exposed skin.
- Ensuring that his protective ensemble remains correctly emplaced.
- Maintaining respiration.
- Controlling bleeding.
- Providing other standard first aid measures
- Transporting the casualty out of the contaminated area.

## B-6. Positioning the Casualty

The first step in any manual carry is to position the casualty to be lifted. If he is conscious, he should be told how he is to be positioned and transported. This helps lessen his fear of movement and to gain his cooperation. It may be necessary to roll the casualty onto his abdomen, or his back, depending upon the position in which he is lying and the particular carry to be used.

*a.* To roll a casualty onto his abdomen, kneel at the casualty's uninjured side.

(1) Place his arms above his head; cross his ankle which is farther from you over the one that is closer to you.

(2) Place your hands on the shoulder which is farther from you; place your other hand in the area of his hip or thigh (Figure B-1).

(3) Roll him gently toward you onto his abdomen (Figure B-2).

*b.* To roll a casualty onto his back, follow the same procedure described in *a* above, except gently roll the casualty onto his back, rather than onto his abdomen.

*Figure B-1. Positioning the casualty.*

*Figure B-2. Rolling casualty onto his abdomen.*

## B-7. Medical Evacuation and Transportation of Casualties

*a.* Medical evacuation of the sick and wounded (with en route medical care) is the responsibility of medical personnel who have been provided special training and equipment. Therefore, unless a good reason for you to transport a casualty arises, wait for some means of medical evacuation to be provided. When the situation is urgent and you are unable to obtain medical assistance or know that no medical evacuation assets are available, you will have to transport the casualty. For this reason, you must know how to transport him without increasing the seriousness of his condition.

*b.* Transporting a casualty by litter (FM 8-10-6) is safer and more comfortable for him than by manual means; it is also easier for you.

Manual transportation, however, may be the only feasible method because of the terrain or the combat situation; or it may be necessary to save a life. In these situations, the casualty should be transferred to a litter as soon as one can be made available or improvised.

## B-8. Manual Carries

Casualties carried by manual means must be carefully and correctly handled, otherwise their injuries may become more serious or possibly fatal. Situation permitting, transport of a casualty should be organized and unhurried. Each movement should be performed as deliberately and gently as possible. Casualties should not be moved before the type and extent of injuries are evaluated and the required first aid is administered. The exception to this occurs when the situation dictates immediate movement for safety purposes (for example, it may be necessary to remove a casualty from a burning vehicle); that is, the situation dictates that the urgency of casualty movement outweighs the need to administer first aid. Manual carries are tiring for the bearers and involve the risk of increasing the severity of the casualty's injury. In some instances, however, they are essential to save the casualty's life. Although manual carries are accomplished by one or two bearers, the two-man carries are used whenever possible. They provide more comfort to the casualty, are less likely to aggravate his injuries, and are also less tiring for the bearers. The distance a casualty can be carried depends on many factors, such as—

- Nature of the casualty's injuries.
- Strength and endurance of the bearer(s).
- Weight of the casualty.
- Obstacles encountered during transport (natural or manmade).
- Type of terrain.

*a. One-man Carries.* These carries should be used when only one bearer is available to transport the casualty.

(1) The *fireman's carry* (Figure B-3) is one of the easiest ways for one individual to carry another. After an unconscious or disabled casualty has been properly positioned, he is raised from the ground, then supported and placed in the carrying position.

(*a*) After rolling the casualty onto his abdomen, straddle him. Extend your hands under his chest and lock them together.

(*b*) Lift the casualty to his knees as you move backward.

(*c*) Continue to move backward, thus straightening the casualty's legs and locking his knees.

(*d*) Walk forward, bringing the casualty to a standing position; tilt him slightly backward to prevent his knees from buckling.

(*e*) As you maintain constant support of the casualty with one arm, free your other arm, quickly grasp his wrist, and raise his arm high. Instantly pass your head under his raised arm, releasing it as you pass under it.

(*f*) Move swiftly to face the casualty and secure your arms around his waist. Immediately place your foot between his feet and spread them apart (approximately 6 to 8 inches).

(*g*) Grasp the casualty's wrist and raise his arm high over your head.

(*h*) Bend down and pull the casualty's arm over and down on your shoulder, bringing his body across your shoulders. At the same time, pass your arm between his legs.

(*i*) Grasp the casualty's wrist with one hand, and place your other hand on your knee for support.

(*j*) Rise with the casualty positioned correctly. Your other hand is free for use.

*Figure B-3. Fireman's carry (Illustrated A—J).*

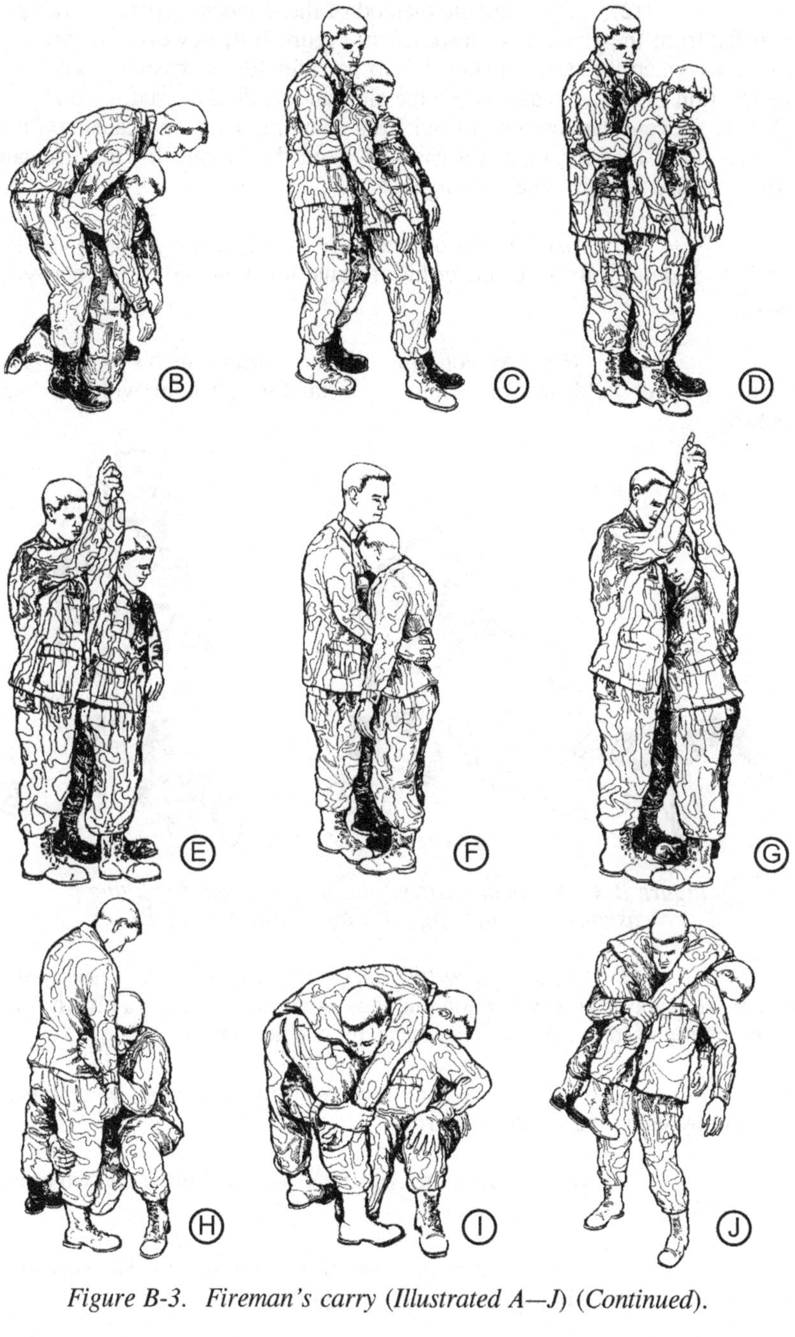


*Figure B-3. Fireman's carry (Illustrated A—J) (Continued).*

(2) The alternate method of the *fireman's carry* for raising a casualty from the ground is illustrated in Figure B-4; however, it should be used only when the bearer believes it to be safer for the casualty because of the location of his wounds. When the alternate method is used, care must be taken to prevent the casualty's head from snapping back and causing a neck injury. The steps for raising a casualty from the ground for the fireman's carry are also used in other one-man carries.

(*a*) Kneel on one knee at the casualty's head and face his feet. Extend your hands under his armpits, down his sides, and across his back.

(*b*) As you rise, lift the casualty to his knees. Then secure a lower hold and raise him to a standing position with his knees locked.

*Figure B-4. Fireman's carry (alternate method) for lifting a casualty to a standing position (Illustrated A—B).*

(3) In the *supporting carry* (Figure B-5), the casualty must be able to walk or at least hop on one leg, using the bearer as a crutch. This carry can be used to assist him as far as he is able to walk or hop.

(*a*) Raise the casualty from the ground to a standing position by using the fireman's carry.

(*b*) Grasp the casualty's wrist and draw his arm around your neck.

(*c*) Place your arm around his waist. The casualty is now able to walk or hop using you as a support.

*Figure B-5. Supporting carry.*

(4) The *arms carry* (Figure B-6) is useful in carrying a casualty for a short distance (up to 50 meters) and for placing him on a litter.

(*a*) Raise or lift the casualty from the ground to a standing position, as in the fireman's carry.

(*b*) Place one arm under the casualty's knees and your other arm around his back.

(*c*) Lift the casualty.

(*d*) Carry the casualty high to lessen fatigue.

*Figure B-6. Arms carry.*

(5) Only a conscious casualty can be transported by the *saddleback carry* (Figure B-7), because he must be able to hold onto the bearer's neck. To use this technique—

(*a*) Raise the casualty to an upright position, as in the fireman's carry.

(*b*) Support the casualty by placing an arm around his waist. Move to the casualty's side. Have the casualty put his arm around your neck and move in front of him with your back to support him.

(*c*) Have the casualty encircle his arms around your neck

(*d*) Stoop, raise him on your back and clasp your hands together beneath his thighs, if possible.

*Figure B-7. Saddleback carry.*

(6) In the *pack-strap carry* (Figure B-8), the casualty's weight rests high on the your back. This makes it easier for you to carry the casualty a moderate distance (50 to 300 meters). To eliminate the possibility of injury to the casualty's arms, you must hold his arms in a palms-down position.

(*a*) Lift the casualty from the ground to a standing position, as in the fireman's carry.

(*b*) Support the casualty with your arms around him and grasp his wrist closer to you.

(*c*) Place his arm over your head and across your shoulders.

(*d*) Move in front of him while still supporting his weight against your back.

(*e*) Grasp his other wrist and place this arm over your shoulder.

(*f*) Bend forward and raise or hoist the casualty as high on your back as possible so that his weight is resting on your back.

**NOTE**

Once the casualty is positioned on the bearer's back, the bearer remains as erect as possible to prevent straining or injuring his back.

*Figure B-8. Pack-strap carry.*

(7) The *pistol-belt carry* (Figure B-9) is the best one-man carry for a long distance (over 300 meters). The casualty is securely supported upon your shoulders by a belt. Both your hands and the casualty's (if conscious) are free for carrying a weapon or equipment, or climbing obstacles. With your hands free and the casualty secured in place, you are also able to creep through shrubs and under low-hanging branches.

(*a*) Link two pistol belts (or three, if necessary) together to form a sling. Place the sling under the casualty's thighs and lower back so that a loop extends from each side.

**NOTE**

If pistol belts are not available for use, other items such as a rifle sling, two cravat bandages, two litter straps, or any other suitable material, which will not cut or bind the casualty may be used.

(*b*) Lie face up between the casualty's outstretched legs. Thrust your arms through the loops and grasp his hands and trouser leg on his injured side.

(*c*) Roll toward the casualty's uninjured side onto your abdomen, bringing him onto your back. Adjust the sling, if necessary.

(*d*) Rise to a kneeling position. The belt will hold the casualty in place.

(*e*) Place one hand on your knee for support and rise to an upright position. (The casualty is supported on your shoulders.)

(*f*) Carry the casualty with your hands free for use in rifle firing, climbing, or surmounting obstacles.

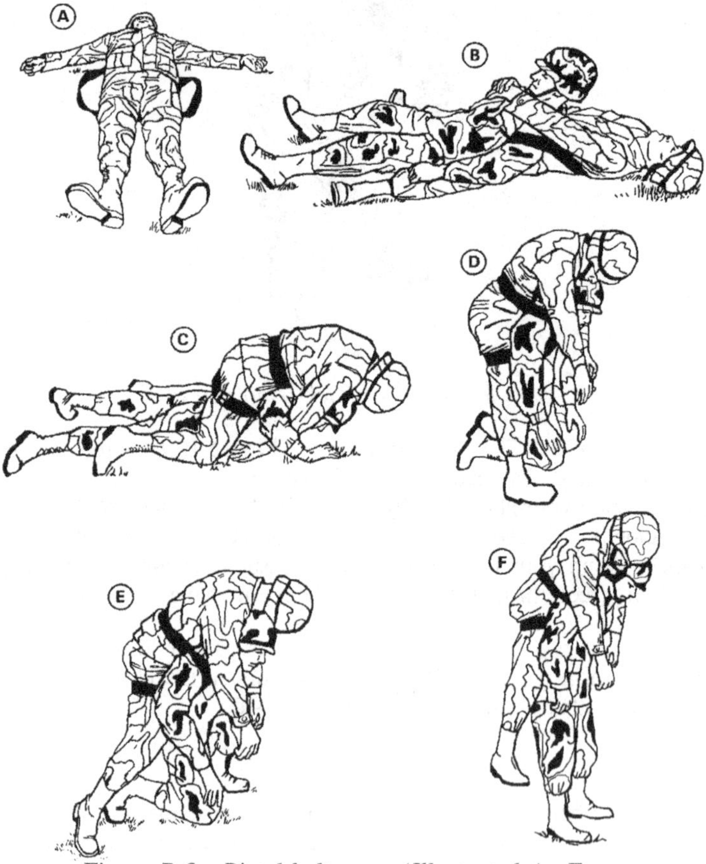


*Figure B-9. Pistol-belt carry (Illustrated A—F).*

(8) The *pistol-belt drag* (Figure B-10), as well as other drags, is generally used for short distances (up to 50 meters). This drag is useful in combat, since both the bearer and the casualty can remain closer to the ground than in any other drags.

(*a*) Extend two pistol belts or similar objects to their full length and join them together to make a continuous loop.

(*b*) Roll the casualty onto his back, as in the fireman's carry.

(*c*) Pass the loop over the casualty's head, and position it across his chest and under his armpits. Then cross the remaining portion of the loop, thus forming a figure eight. Keep tension on the belts so they do not come unhooked.

(*d*) Lie on your side facing the casualty.

(*e*) Slip the loop over your head and turn onto your abdomen. This enables you to drag the casualty as you crawl.

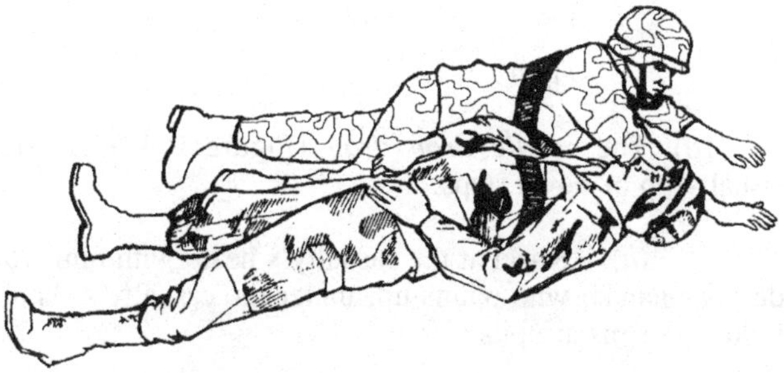

*Figure B-10. Pistol-belt drag.*

(9) The *neck drag* (Figure B-11) is useful in combat because the bearer can transport the casualty as he creeps behind a low wall or shrubbery, under a vehicle, or through a culvert. If the casualty is unconscious, his head must be protected from the ground. The neck drag cannot be used if the casualty has a broken arm.

**NOTE**

If the casualty is conscious, he may clasp his hands together around your neck.

(*a*) Tie the casualty's hands together at the wrists.

(*b*) Straddle the casualty in a kneeling face-to-face position.

(*c*) Loop the casualty's tied hands over and around your neck.

(*d*) Crawl forward dragging the casualty with you.

**NOTE**

If the casualty is unconscious, protect his head from the ground.

*Figure B-11. Neck drag.*

(10) The *cradle drop drag* (Figure B-12) is effective in moving a casualty up or down steps.

(*a*) Kneel at the casualty's head (with him lying on his back). Slide your hands, with palms up, under the casualty's shoulders and get a firm hold under his armpits.

(*b*) Rise (partially), supporting the casualty's head on one of your forearms. (You may bring your elbows together and let the casualty's head rest on both of your forearms.)

(*c*) Rise and drag the casualty backward. (The casualty is in a semisitting position.)

(*d*) Back down the steps, supporting the casualty's head and body and letting his hips and legs drop from step to step.

**NOTE**

If the casualty needs to be moved up the steps, you should back up the steps, using the same procedure.

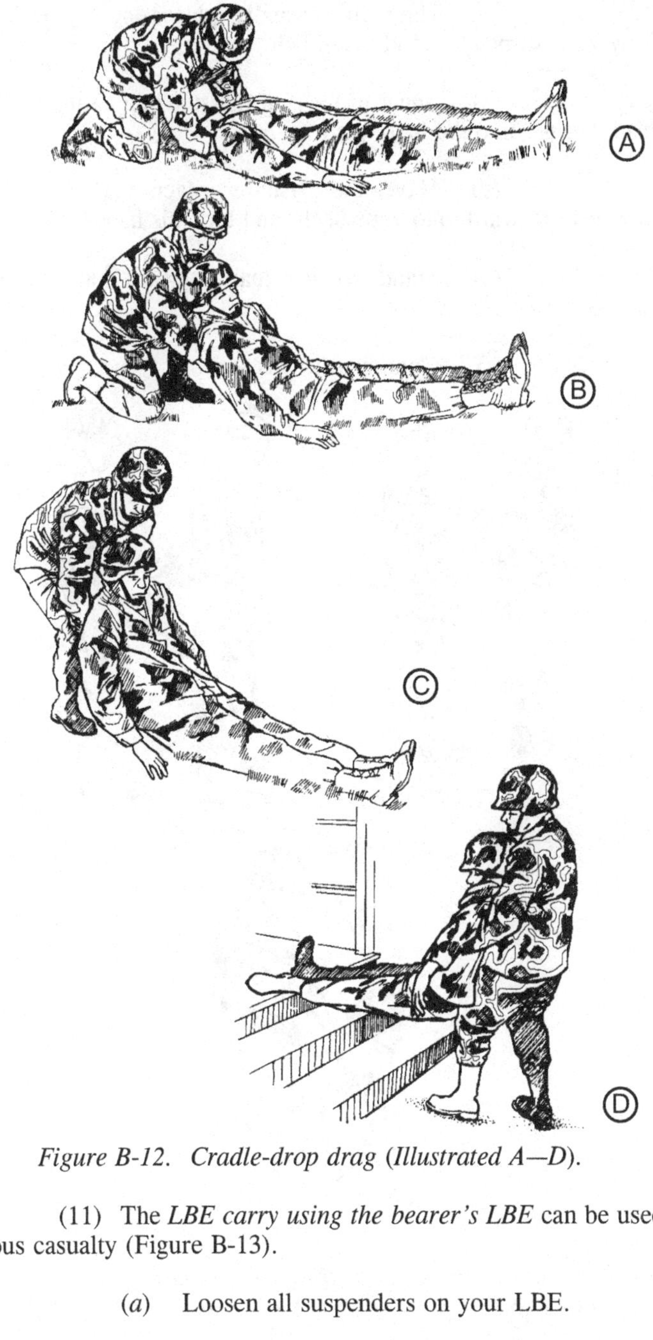


*Figure B-12. Cradle-drop drag (Illustrated A—D).*

(11) The *LBE carry using the bearer's LBE* can be used with a conscious casualty (Figure B-13).

(*a*) Loosen all suspenders on your LBE.

(*b*) Have the casualty place one leg into the loop formed by your suspenders and pistol belt.

(*c*) Squat in front of the standing casualty. Have him place his other leg into the loop, also.

(*d*) Have the casualty place his arms over your shoulders, lean forward onto your back, and lock his hands together.

(*e*) Stand up and lean forward into a comfortable position.

(*f*) Continue the mission.

*Figure B-13. Load bearing equipment carry using bearer's LBE (conscious casualty) (Illustrated A—F).*

*Figure B-13. Load bearing equipment carry using bearer's LBE (conscious casualty) (Illustrated A—F) (Continued).*

(12) The *LBE carry using the bearer's LBE* can be used with an unconscious casualty or one who cannot stand (Figure B-14).

(*a*) Position the casualty on the flat of his back.

(*b*) Remove your LBE and loosen all suspender straps.

(*c*) Lift the casualty's leg and place it through the loop formed by your suspenders and pistol belt. Then place the other leg through the same loop. The LBE is moved up until the pistol belt is behind the casualty's thighs.

(*d*) Lay between the casualty's legs; work your arms through the LBE suspenders.

(*e*) Grasp the casualty's hand (on the injured side), and roll the casualty (on his uninjured side) onto your back.

(*f*) Rise to one knee and then push into a standing position.

(*g*) Bring the casualty's arms over your shoulders. Grasp his hands and secure them if the casualty is unconscious. If the casualty is conscious, have him lock his hands in front if he is able to do so.

(*h*) Lean forward into a comfortable position and continue the mission.

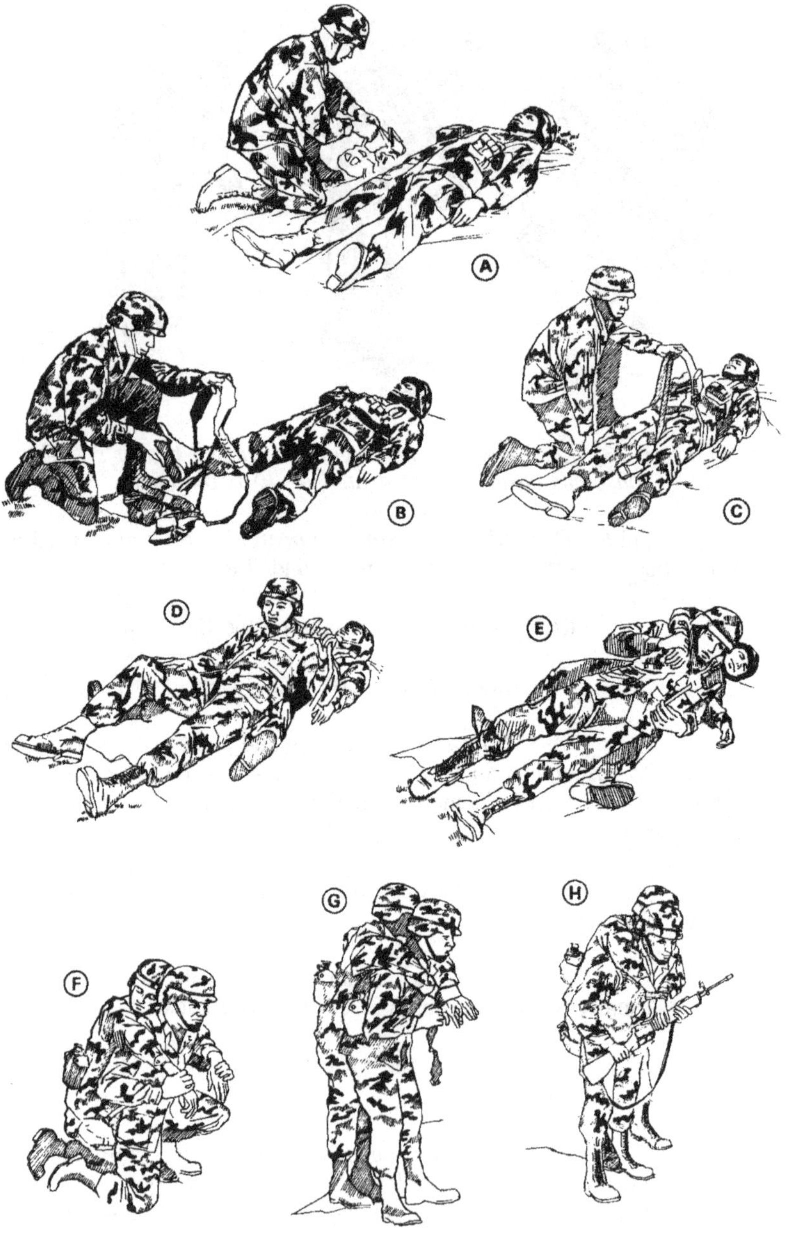


*Figure B-14. Load bearing equipment carry using bearer's LBE (unconscious casualty or one that cannot stand) (Illustrated A—H).*

(13) The *LBE carry using the casualty's LBE* (Figure B-15) can be used with a conscious or unconscious casualty.

(*a*) Position the casualty on his back with his LBE on.

(*b*) Loosen the casualty's two front suspenders.

(*c*) Position yourself between the casualty's legs, and slip your arms into the casualty's two front suspenders (up to your shoulders).

(*d*) Work his arms out of his LBE suspenders.

(*e*) Grasp the casualty's hand (on the injured side), and roll him (on his uninjured side) onto your back.

(*f*) Rise to one knee, then into a standing position.

(*g*) Grasp the casualty's hands and secure them, if the casualty is unconscious. Have the casualty lock his hands in front of you, if he is conscious.

(*h*) Lean forward into a comfortable position and continue the mission.

*Figure B-15. Load bearing equipment carry using casualty's LBE (Illustrated A—G).*

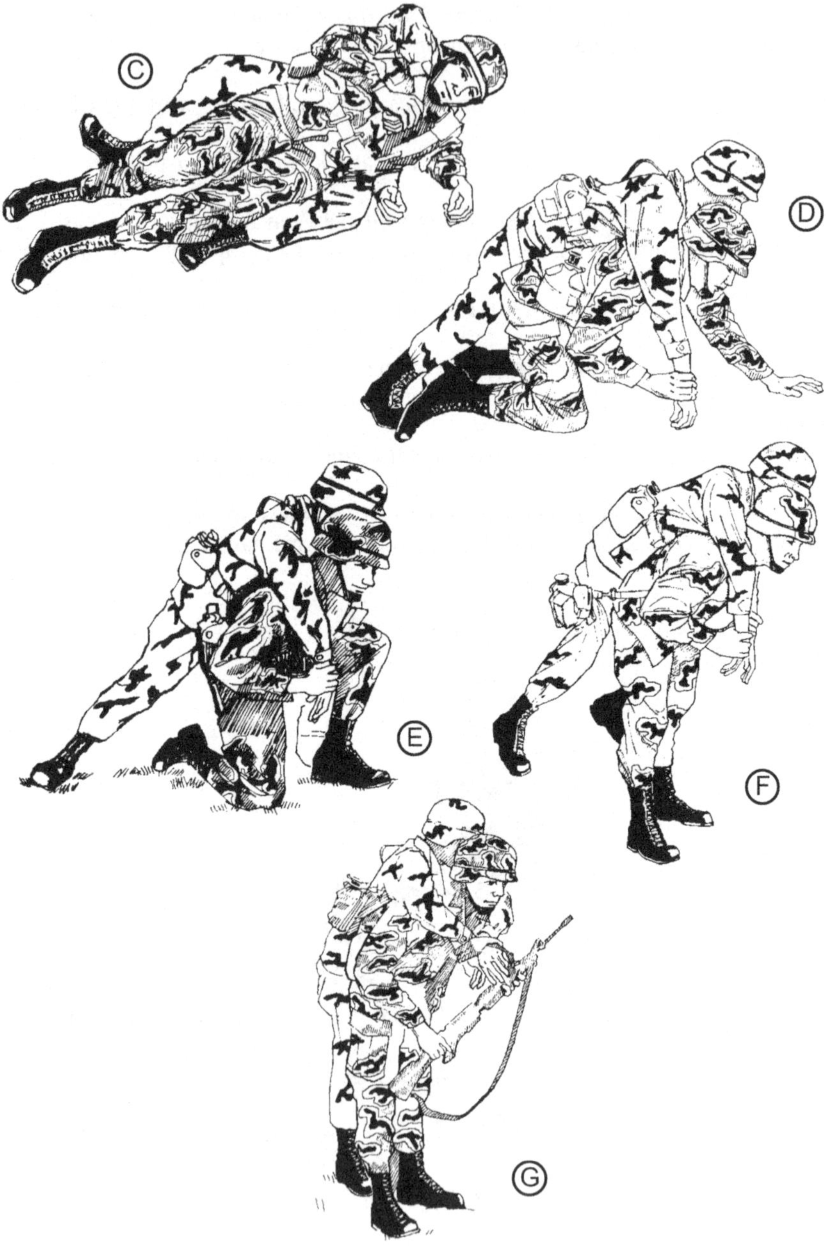


*Figure B-15. Load bearing equipment carry using casualty's LBE (Illustrated A—G) (Continued).*

*b*. *Two-man Carries*. These carries should be used whenever possible. They provide more casualty comfort, are less likely to aggravate injuries, and are less tiring for the bearers. Five different two-man carries can be used.

(1) The *two-man support carry* (Figure B-16) can be used in transporting either conscious or unconscious casualties. If the casualty is taller than the bearers, it may be necessary for the bearers to lift the casualty's legs and let them rest on their forearms. The bearers—

(*a*) Help the casualty to his feet and support him with their arms around his waist.

(*b*) Grasp the casualty's wrists and draw his arms around their necks.

*Figure B-16. Two-man supporting carry.*

(2) The *two-man arms carry* (Figure B-17) is useful in carrying a casualty for a moderate distance (50 to 300 meters) and placing him on a litter. To lessen fatigue, the bearers should carry the casualty high and as close to their chests as possible. In extreme emergencies when there is no time to obtain a spine board, this carry is the safest one for transporting a casualty with a back injury. If possible, two additional bearers should be used to keep the casualty's head and legs in alignment with his body. The bearers—

(*a*) Kneel at one side of the casualty; then they place their arms beneath the casualty's back, waist, hips, and knees.

(*b*) Lift the casualty while rising to their knees.

(*c*) Turn the casualty toward their chests, while rising to a standing position. Carry the casualty high to lessen fatigue.

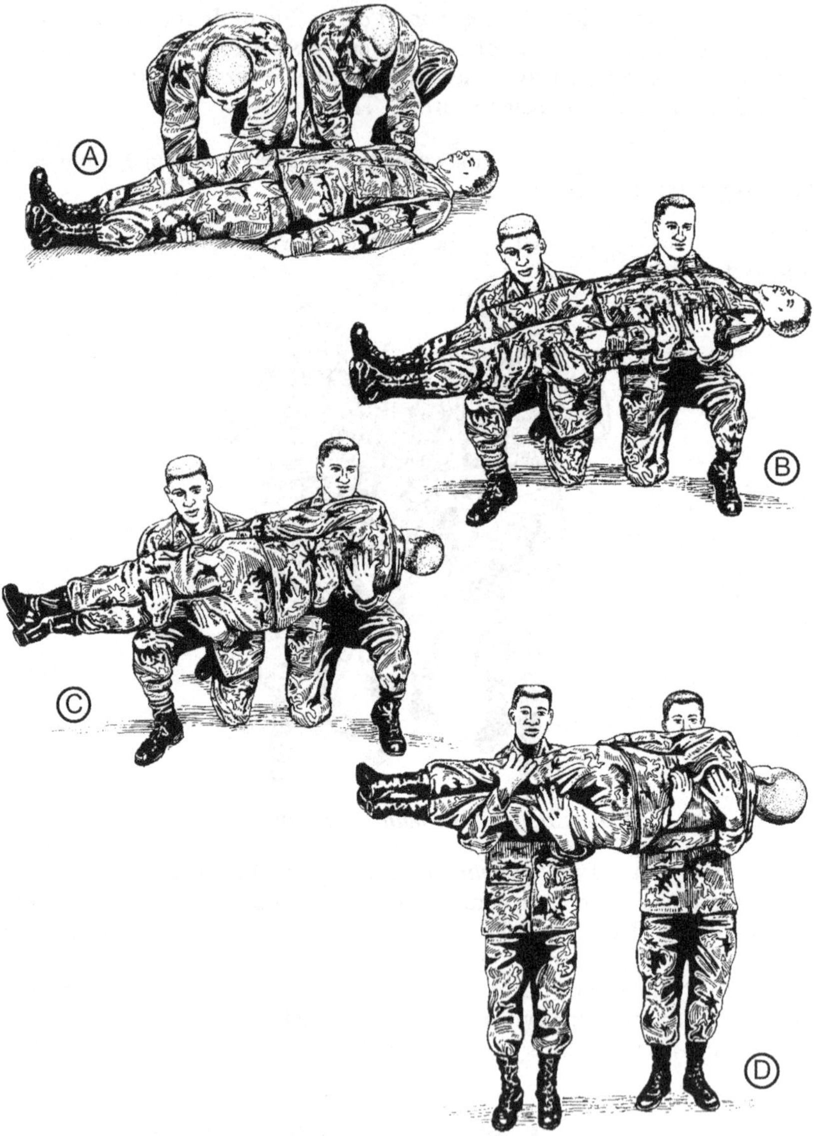


*Figure B-17. Two-man arms carry (Illustrated A—D).*

(3) The *two-man fore-and aft-carry* (Figure B-18) is a useful *two-man carry* for transporting a casualty for a long distance (over 300 meters). The taller of the two bearers should position himself at the casualty's head. By altering this carry so that both bearers face the casualty, it is useful for placing a casualty on a litter.

(*a*) The shorter bearer spreads the casualty's legs and kneels between them with his back to the casualty. He positions his hands behind the casualty's knees. The other bearer kneels at the casualty's head, slides his hands under the arms, across the chest, and locks his hands together.

(*b*) The two bearers rise together, lifting the casualty.

*Figure B-18. Two-man fore-and-aft carry (Illustrated A—B).*

(4) Only a conscious casualty can be transported with the *four-hand seat carry* (Figure B-19) because he must help support himself by placing his arms around the bearers' shoulders. This carry is especially useful in transporting a casualty with a head or foot injury for a moderate distance (50 to 300 meters). It is also useful for placing a casualty on a litter.

(*a*) Each bearer grasps one of his wrists and one of the other bearer's wrists, thus forming a packsaddle.

(*b*) The two bearers lower themselves sufficiently for the casualty to sit on the packsaddle; then, they have the casualty place his arms around their shoulders for support. The bearers then rise to an upright position.

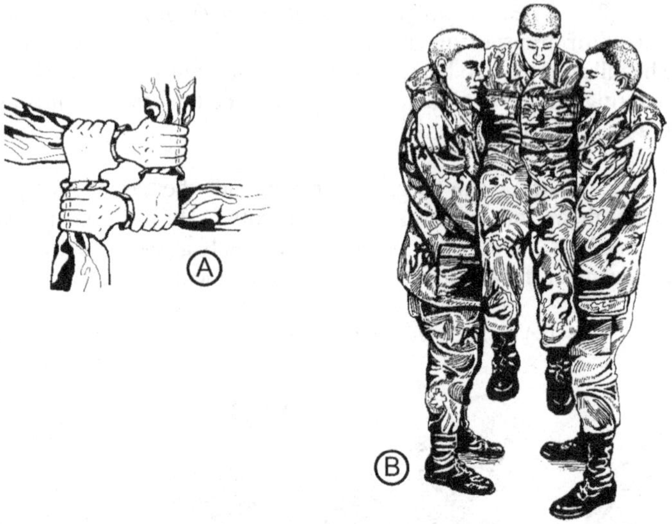

Figure B-19. Four-hand seat carry (Illustrated A—B).

(5) The *two-hand seat carry* (Figure B-20) is used when carrying a casualty for a short distance or for placing him on a litter. With the casualty lying on his back, a bearer kneels on each side of the casualty at his hips. Each bearer passes his arms under the casualty's thighs and back, and grasps the other bearer's wrists. The bearers rise lifting the casualty.

Figure B-20. Two-hand seat carry (Illustrated A—B).

## B-9. Improvised Litters

Two men can support or carry a casualty without equipment for only short distances. By using available materials to improvise equipment, the casualty can be transported greater distances by two or more rescuers.

*a.* There are times when a casualty may have to be moved and a standard litter is not available. The distance may be too great for manual carries or the casualty may have an injury (such as a fractured neck, back, hip, or thigh) that would be aggravated by manual transportation. In these situations, litters can be improvised from materials at hand. Improvised litters must be as well constructed as possible to avoid risk of dropping or further injuring the casualty. Improvised litters are emergency measures and must be replaced by standard litters at the first opportunity.

*b.* Many different types of litters can be improvised, depending upon the materials available. A satisfactory litter can be made by securing poles inside such items as a blanket, poncho, shelter half, tarpaulin, mattress cover, jacket, shirt, or bed ticks, bags, and sacks (Figure B-18). Poles can be improvised from strong branches, tent supports, skis, lengths of pipe or other objects. If objects for improvising poles are not available, a blanket, poncho, or similar item can be rolled from both sides toward the center so the rolls can be gripped for carrying a patient. Most flat-surface objects of suitable size can be used as litters. Such objects include doors, boards, window shutters, benches, ladders, cots, and chairs. If possible, these objects should be padded for the casualty's comfort.

(1) To improvise a litter using a blanket and poles (Figure B-21), the following steps should be used.

*Figure B-21. Litter made with blanket and poles.*

(*a*) Open the blanket and lay one pole lengthwise across the center; then fold the blanket over the pole.

(*b*) Place the second pole across the center of the folded blanket.

(*c*) Fold the free edges of the blanket over the second pole and across the first pole.

(2) To improvise a litter using shirts or jackets (Figure B-22), button the shirt or jacket and turn it inside out, leaving the sleeves inside, (more than one shirt or jacket may be required), then pass the pole through the sleeves.

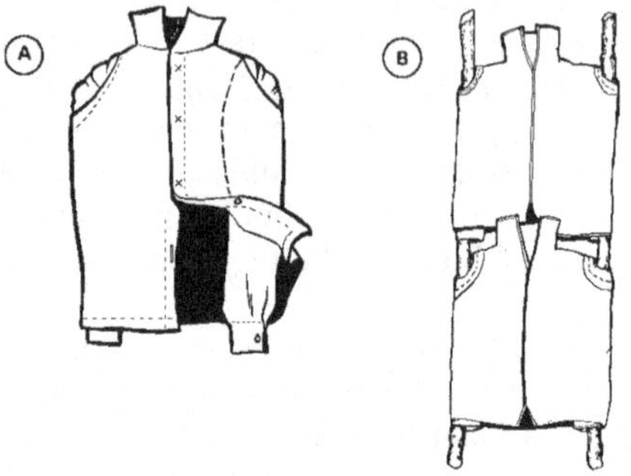


*Figure B-22. Litter improvised from jackets and poles (Illustrated A—B).*

(3) To improvise a litter from bed sacks and poles (Figure B-23), rip open the corners of bed ticks, bags, or sacks; then pass the poles through them.

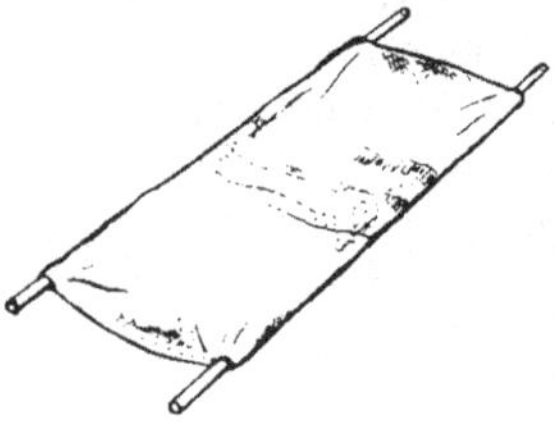

*Figure B-23. Litter improvised from bed sacks and poles.*

(4) If no poles are available, roll a blanket, shelter half, tarpaulin, or similar item from both sides toward the center (Figure B-24). Grip the rolls to carry the casualty.

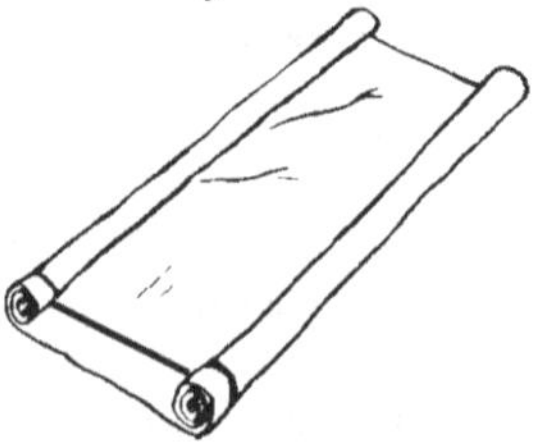

*Figure B-24. Rolled blanket used as a litter.*

*c.* Any of the appropriate carries may be used to place a casualty on a litter. These carries are:

- The one-man arms carry (Figure B-6).
- The two-man arms carry (Figure B-17).
- The two-man fore-and-aft carry (Figure B-18).
- The two-hand seat carry (Figure B-20).
- The four-hand seat carry (Figure B-19).

**WARNING**

**Unless there is an immediate life-threatening situation (such as fire, explosion), DO NOT move a casualty with a suspected back or neck injury. Seek medical personnel for guidance on how to transport.**

*d.* Either two or four service members (head/foot) may be used to lift a litter. To lift the litter, follow the procedure below.

(1) Raise the litter at the same time as the other carriers/ bearers.

(2) Keep the casualty as level as possible.

**NOTE**

Use caution when transporting on a sloping incline/hill.

GLOSSARY

# ACRONYMS, ABBREVIATIONS, AND DEFINITIONS

**AC** hydrogen cyanide
**AFMAN** Air Force Manual
**AOC** area of concentration
**AR** Army regulation
**ATM** advanced trauma management
**ATNAA** Antidote Treatment, Nerve Agent, Autoinjector
**attn** attention

**BDO** battle dress overgarment
**BDU** battle dress uniform
**BZ** anticholinergic drugs

**C** Celsius
**CANA** Convulsant Antidote for Nerve Agent
**CASEVAC** casualty evacuation
**cc** cubic centimeter
**CG** phosgene
**CHS** combat health support
**CK** cyanogen chloride
**Cl** chlorine
**CLS** Combat Lifesaver
**CNS** central nervous system
**$CO_2$** carbon dioxide
**COSR** combat and operational stress reactions
**CSR** combat stress reaction
**CTA** common table of allowance
**CX** phosgene oxime

**DA** Department of the Army
**DD** Department of Defense
**DM** diphenylaminochloroarsine (adamsite)
**DNBI** disease and nonbattle injury
**DOD** Department of Defense
**DP** diphosgene
**DS** direct support

**EMT** emergency medical treatment

**F** Fahrenheit
**FM** field manual

**H** mustard

**HD** mustard
**HM** Hospital Corpsman
**HN** nitrogen mustard
**HSS** health service support

**IPE** individual protective equipment
**IV** intravenous

**JSLIST** Joint Services Light Weight Integrated Suit Technology

**L** lewisite
**lasers** laser means Light Amplification by Stimulated Emission of Radiation and sources include range finders, weapons/guidance, communication systems, and weapons simulations such as MILES [Multiple Integrated Laser Engagement System].
**LBE** load bearing equipment
**LX** lewisite and mustard

**MCRP** Marine Corps Reference Publication
**MILES** Multiple Integrated Laser Engagement System
**ml** milliliter
**MOPP** mission-oriented protective posture
**MOS** military occupational specialty
**MTF** medical treatment facility

**NAPP** Nerve Agent Pyridostigmine Pretreatment
**NATO** North Atlantic Treaty Organization
**NBC** nuclear, biological, and chemical
**NCO** noncommissioned officer
**NTRP** Navy Tactical Reference Publication

**occlusive dressing** air tight transparent dressing used to seal and cover wounds
**oz** ounce

**PAM** pamphlet
**PS** chloropicrin
**PTSD** post-traumatic stress disorder

**QSTAG** Quadripartite Standardization Agreement

**SOP** standing operating procedure
**STANAG** standardization agreement
**STP** soldier training publication

**2 PAM Cl** pralidoxime chloride
**TB MED** technical bulletin medical
**TM** technical manual
**TSOP** tactical standing operating procedure

**US** United States

**WP** white phosphorus

# REFERENCES

## DOCUMENTS NEEDED

These documents must be available to the intended users of this publication.

## NATO STANAGs

These agreements are available on request using DD Form 1425 from Standardization Document Order Desk, 700 Robin Avenue, Building 4, Section D, Philadelphia, Pennsylvania 19111-5094.

2122. *Medical Training in First Aid, Basic Hygiene and Emergency Care.* 10 December 1975.
2126. *First Aid Kits and Emergency Medical Care Kits.* 27 September 1983.
2358. *First Aid and Hygiene Training in NBC Operations.* 3 March 1989.
2871. *First Aid Material for Chemical Injuries.* 10 March 1986.

## ABCA QSTAGs

These agreements are available on request using DD Form 1425 from Standardization Document Order Desk, 700 Robin Avenue, Building 4, Section D, Philadelphia, Pennsylvania 19111-5094.

535. *Medical Training in First Aid, Basic Hygiene and Emergency Care.* 12 November 1979.

### Joint and Multiservice Publications

FM 21-10. *Field Hygiene and Sanitation.* MCRP 4-11.1D. 21 June 2000.

### Army Publications

AR 350-41. *Training In Units.* 19 March 1993.
DA PAM 350-59. *Army Correspondence Course Program Catalog.* 26 October 2001.
FM 3-4. *NBC Protection.* FMFM 11-9. 29 May 1992 (Reprinted with basic including Change 1, 28 October 1992; Change 2, 26 February 1996.)
FM 3-5. *NBC Decontamination.* MCWP 3-37.3. 28 July 2000. (Change 1, 31 January 2002.)

FM 3-100. *Chemical Operations Principles and Fundamentals.* MCWP 3-3.7.1. 8 May 1996.
FM 4-02.33 (8-33). *Control of Communicable Diseases Manual* (17th Edition). 3 January 2000.
FM 8-10-6 (4-02.2). *Medical Evacuation in a Theater of Operations—Tactics, Techniques, and Procedures.* 14 April 2000.
FM 4-02.7 (8-10-7). *Health Service Support in a Nuclear, Biological, and Chemical Environment.* 1 October 2002.
FM 8-284 (4-02.284). *Treatment of Biological Warfare Agent Casualties.* NAVMED P-5042; AFMAN (I) 44-156; MCRP 4-11.1C. 17 July 2000. (Change 1, 8 July 2002.)
FM 8-285 (4-02.285). *Treatment of Chemical Agent Casualties and Conventional Military Chemical Injuries.* NAVMED P-5041; AFJMAN 44-149; FMFM 11-11. 22 December 1995.
FM 22-51 (4-02.22). *Leaders' Manual for Combat Stress Control.* 29 September 1994.
CTA 8-100. *Army Medical Department Expendable/Durable Items.* 31 August 1994.
CTA 50-900. *Clothing and Individual Equipment.* 1 September 1994.
STP 21-1-SMCT. *Soldier's Manual of Common Tasks Skill Level I.* 1 October 2001.

### Department of Defense Forms

DD Form 1425. *Specifications and Standards Requisition.* March 1986.

## READINGS RECOMMENDED

These readings contain relevant supplemental information.

### Joint and Multiservice Publications

FM 8-9 (4-02.11). *NATO Handbook on the Medical Aspects of NBC Defensive Operations AMEDP-6 (B), Part I—Nuclear, Part II—Biological, Part III—Chemical.* NAVMED P-5059; AFJMAN 44-151V1V2V3. 1 February 1996.
TB MED 81. *Cold Injury.* NAVMED P-5052-29; AFP 161-11. 30 September 1976.
TB MED 507. *Occupational and Environmental Health Prevention, Treatment and Control of Heat Injury.* NAVMED P-5052-5; AFP 160-1. 25 July 1980.

**Army Publications**

AR 310-25. *Dictionary of United States Army Terms (Short Title: AD)*. 15 October 1983. (Reprinted with basic including Change 1, 21 May 1986.)

AR 310-50. *Authorized Abbreviations and Brevity Codes*. 15 November 1985.

TM 3-4230-216-10. *Operator's Manual for Decontaminating Kit, Skin: M258A1 (NSN 4230-01-101-3984) and Training Aid, Skin Decontaminating: M58A1 (6910-01-101-1768)*. 17 May 1985. (Change 1, 22 January 1997.)

# INDEX

References are to paragraph numbers except where specified otherwise.

TRAINING CIRCULAR
No. 21-3

HEADQUARTERS
DEPARTMENT OF THE ARMY
Washington, DC, 17 March 1986

# SOLDIER'S HANDBOOK FOR INDIVIDUAL OPERATIONS AND SURVIVAL IN COLD-WEATHER AREAS

## TABLE OF CONTENTS



---

*** This publication supersedes TC 21-3, 30 September 1974.**

**CHAPTER 9. LAND NAVIGATION**



**CHAPTER 10. SURVIVAL**

# PREFACE

This handbook is an aid for soldiers in cold climates. However, expertise in dealing with cold-weather operations cannot be gained just from reading this handbook. To develop successful cold-weather field techniques, application training and operational experience are necessary. Additional information is available in FM 31-70, FM 31-71, FM 90-11 (TBP), and FM 9-207.

Military history of operations in northern areas shows that man's success or failure is measured by his regard for the dominant characteristics of a cold-weather environment. For a soldier to become a cold-weather expert, he must learn the techniques and rules needed to fight and live in a cold-weather climate.

The proponent of this publication is HQ TRADOC. Submit changes for improving this publication on DA Form 2028 (Recommended Changes to Publications and Blank Forms) and forward it to Commandant, US Army Infantry School, ATTN: ATSH-B, Fort Benning, Georgia 31905-5593.

Unless otherwise stated, whenever the masculine gender is used, both men and women are included.

# CHAPTER 1

# Introduction to Cold-Weather Operations

***"Soldiers in pairs look after each other; be faithful to your mate; be loyal to your leaders."***

***Anonymous***

## 1-1. COLD CONDITIONS

### COLD-WET

Cold-wet conditions occur when temperatures are near freezing and high and low temperatures cause alternate freezing and thawing. These conditions can be accompanied by wet snow and rain, causing the ground to become slushy and muddy. Soldiers need clothing that has a waterproof or water-repellent, wind-resistant outer layer, and an inner layer with insulation that provide protection in moderately cold weather (above 14°F). **Waterproof footwear is essential.**

### COLD-DRY

Cold-dry conditions occur when average temperatures are lower than 14°F. The ground is usually frozen and the snow dry. Low temperatures plus wind increase the need for protection of the entire body. For these conditions, soldiers need clothing that will provide insulation for a wind-chill factor of -80°F. For temperatures below 14°F, white vapor barrier (VB) boots effective down to temperatures of -50°F must be worn.

### WIND CHILL

Wind chill is a measure of the combined effects of wind and temperature. Temperature alone does not give a true indication of the impact of the cold. In order to effectively gauge the difference between temperature and the impact of the cold, a wind-chill scale must be used. The chart in Appendix A is a simple and practical guide showing the conditions under which cold weather is dangerous and when exposed flesh is likely to freeze. Soldiers must be familiar with the wind-chill chart to recognize these dangers.

## 1-2. EFFECTS OF COLD WEATHER ON MILITARY OPERATIONS

Cold has a numbing effect on skin, which hinders task performance. This must be considered when accomplishing such routine tasks as vehicle maintenance, and making and striking bivouac areas (FM 9-207).

When conditions become extreme and the temperature drops, the problem of survival becomes critical. Under these circumstances soldiers may tend to withdraw emotionally, not wanting to leave the warmth and protection of their shelter or tent. The remedy is physical activity. Try to remain alert and active. Execution of all military duties and the fast and proper performance of group tasks are vital.

### COCOON-LIKE EXISTENCE

Many soldiers, when wearing several layers of clothing and with their heads covered, tend to withdraw within themselves and assume a cocoon-like existence. When so clothed, many soldiers' hearing and field of vision are restricted; they tend not to notice their surroundings. Their thinking and reasoning become sluggish.

### INDIVIDUAL AND GROUP HIBERNATION

This process is characterized by individuals seeking the comfort of sleeping bags, and by the group remaining in tents or other shelter. Guard and security measures may be neglected, jeopardizing the security and safety of the unit.

## 1-3. OVERCOMING THE COLD

The body must be protected. It must be kept clean, dry, and warm, with normal body processes maintained. Rest and nourishment are vital. The right approach to cold-weather living will keep you healthy and get you through challenging times. There are four basic rules to follow:

### KEEP IN SHAPE

Cold-weather clothing and equipment are heavy, and add weight to your normal equipment load. This hinders movement in snow and uses too much energy. The importance of being in excellent physical condition cannot be overemphasized. Your unit must have a tough, challenging, demanding physical training program.

### DRINK PLENTY OF WATER

Water may be hard to get, so you may drink less than you need or only when you are thirsty. This will not give you the water needed to avoid dehydration. Drink plenty of water (at least 3.5 quarts per day) to avoid dehydration and fatigue. DO NOT eat snow as a water substitute; it will lower your body's core temperature.

### EAT TO KEEP FIT

Regular, nutritious hot food is needed for top performance. You must eat, even when you are not hungry.

### KEEP A GOOD ATTITUDE

You will meet many new challenges, but none that cannot be overcome. Your attitude toward the cold will reflect those of your leaders.

## 1-4. POSITIVE LEADERSHIP AND THE RIGHT ATTITUDE

Most soldiers, after 3 to 4 weeks of cold-weather training, will be able to cope with the environment. Some soldiers may not be as skilled as others, but most will succeed.

Set an example for other soldiers. Winter conditions can be frightening, trying to meet challenges you have never met before. Any mistake could make you a casualty. Aggressive leadership actions that overcome the challenges of the environment are essential to mission accomplishment.

There are two enemies to overcome in northern operations: the enemy soldier and the cold. The first step in defeating either enemy is having the right attitude. The cold can defeat you psychologically if you are not aware of the symptoms of a poorly prepared unit. The following are suggestions on how to defeat the cold:

- If you become depressed or moody, or do not want to talk, force yourself to get involved with others. Do not leave out the weaker soldier.
- If you find it hard to remember things you have been taught, review orders and drills.
- Remind soldiers that their jobs are to close with, engage, and gut the enemy, and that weapons and equipment must be kept in fighting order. Operations in a northern environment can easily become a camping trip with serious consequences. You must avoid this common ambush.
- Cold weather may slow down the following of an order, but it is not an excuse for not doing it.
- Cold weather may make tasks harder and take longer to do, but it will not make them impossible. With knowledge, proper equipment, and proper training you can overcome the cold and be successful in combat.

# CHAPTER 2

# Personal Clothing and Equipment

***"The freezing German troops were reduced to removing clothes from enemy corpses, improvising strawboots, and taking other emergency measures."***

***Russia, 1942***

## 2-1. INDIVIDUAL CLOTHING

Soldiers should understand the design principles of the military cold-weather clothing system. These principles are: insulate, layer, and ventilate. There are four ways to make these principles work for you: keep it Clean, avoid Overheating, wear it Loose, and keep it Dry. By remembering the "buzz word" COLD, you can quickly check your clothing.

### INSULATE

Insulation material reduces the amount of heat loss. By regulating the amount of insulation, an individual can regulate the amount of heat lost or retained.

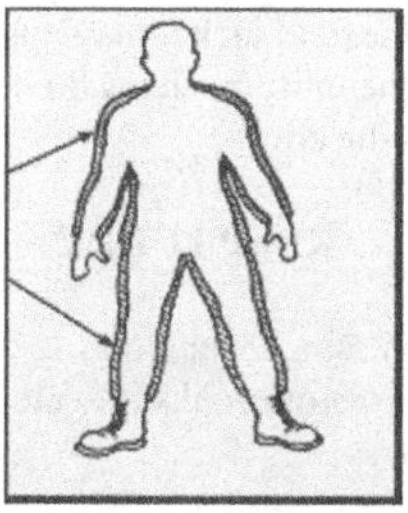

**Figure 2-1. Insulation.**

### LAYER

Several layers of clothing provide more insulation and movement than one heavy garment, even if the heavy garment is as thick as the combined layers. Air pockets trapped between the layers provide insulation and warmth.

### VENTILATE

Ventilation helps maintain a comfortable body temperature by allowing excess heat and moisture to escape.

**Figure 2-2. C-0-L-D.**

### KEEP CLOTHING CLEAN

Dirty clothes are cold clothes. Dirt and grease clog the air spaces in clothing and reduce insulation.

### AVOID OVERHEATING

Select comfortable clothing and do not overdress. In cold weather, it is better to be a little cool rather than risk being overheated.

### WEAR IT LOOSE

All items of the cold-weather uniform are large enough to allow wearing of the appropriate number of layers. Your field jacket may appear too large without all of the layers designed to fit under it. If the uniform items do not fit loosely, the insulation that keeps you warm will be greatly reduced.

### KEEP IT DRY

It is important to keep clothing dry, since wet clothing conducts heat away from the body more quickly than dry clothing. Moisture soaks into clothing from two directions: the inside and the outside.

From the outside you must guard against melting snow; from the inside you must guard against sweating. Brush snow and frost from clothing before entering any heated shelter or vehicle, and avoid overheating.

## 2-2. COLD-WET VERSUS COLD-DRY

As mentioned in Chapter 1, temperatures that are above 14°F (-10° C) are considered cold-wet conditions; temperatures consistently below 14°F will cause cold-dry conditions. Cold-wet conditions usually involve a freeze/thaw cycle, often accompanied by rain or wet snow. This type of weather is hard to dress for, since each soldier must battle the elements to keep his clothing dry. During cold conditions, the ground is usually frozen and the snow is dry. Athough the temperatures during the cold-dry conditions may be lower than during cold-wet conditions, it is easier to stay warm.

| COLD-WET | | COLD-DRY |
|---|---|---|
| | OR | |
| DOWN TO 14°F | | BELOW 14°F |

**Figure 2-3. Cold-wet versus cold-dry.**

## 2-3. THE COLD-WEATHER UNIFORM

There are many layers of clothing that must be worn in a cold weather environment. The first layer of clothing is the T-shirt and shorts. The second layer is a set of thermal underwear, which consists of an undershirt and underpants. You should wear only the clothing needed to keep comfortable and warm.

The scissors-type suspenders must be the last thing next to the outer layer, or else the layers cannot be tucked inside the trousers. The underpants and all succeeding layers of trousers are supported from the metal hooks on the suspenders. The suspenders allow the underpants and trousers to be worn loosely at the waist so that circulation and ventilation are not restricted.

Wear the wool shirt outside the trousers, which provides another method of ventilation. You can increase the waist size of this shirt by removing the stitching in each dart at the sides of the shirt.

Wear the field trousers as an outer garment in the cold-wet environment. In the cold-dry environment, wear the nylon liners inside the field trousers during periods of inactivity. On the inside of each cargo pocket on the trousers is a tape that runs through a hole on the side of the pocket. The tape ties around the thighs and keeps the cargo in the pocket from rubbing and irritating the legs. For warmth and to prevent entry of snow, tie the drawstrings around your ankles outside the boots into the boot lacing, and tuck the bowknots under the trousers.

The field jacket is worn with the liner in both cold-wet and cold-dry areas. Wear the coat and liner directly over the wool shirt. Remove the liner in mild weather or when activity increases.

In a cold-dry environment the parka is worn with a liner. The waist is adjustable for ventilation and warmth with a drawcord, and the skirt has a "split-tail" that can be tied tightly around each leg for warmth. During a period of severe cold, when activity is less, wear the parka with liner over your field jacket with liner. Remember: do not overdress or you will sweat.

Do not dry nylon on drying lines too close to the heat; too much heat will melt the nylon, destroying its insulating value.

Keep the head covered to stay warm; a warm head even warms the feet. Wear the cold-weather cap or balaclava under your helmet or under the hood to your field jacket. When wearing the cold-weather cap as an outer headpiece, the lower flap may be folded up over the top and fastened with hook-pile tape straps in front. In a cold-dry environment, wear the winter hood over the cap or the helmet. Adjust the hood by using the drawstrings on either side of the hood and the elastic strip at the back. To provide minimum protection for the face from the wind, bend the wire in the fur cuff of the hood so that you have a small opening for your face. This allows visibility but less protection.

The feet are hard to keep warm and dry, so protect them. The insulated boots (black for cold-wet conditions and white for cold-dry conditions) are warm and waterproof. The white insulated boot is the best cold-weather footwear today. When you are active, your feet will sweat and, because of the waterproof rubber lining, the moisture will remain in the socks or in the bottom of the boots. The feet, even if damp, will stay warm because the heat given off by your body will remain inside the boots. If the boots are worn for long periods, the feet become white and wrinkled when you remove the boots. This is no cause for alarm. Drying, massaging with foot powder, warming and putting on dry socks will correct this. If you are inactive for long periods, your feet may become cold. Exercises such as knee bends, stamping your feet, running in place, or wiggling your toes will make them warm. Elevate the feet whenever possible to aid blood circulation. Don't wear socks that are too tight, and wear only one pair of cushion-sole socks at a time. Change your socks at least twice a day. Most important, stay active.

Care of your boots is important. Wash boots inside and out with soapy water at least once per month or as required by unit SOP you want your feet to keep warm, you must keep the insulation dry. Inspect your shoes often and repair any holes or punctures immediately. Temporary repairs can be made by patching the hole with any kind of tape; even chewing gum can be used. Seal the hole quickly to prevent the insulation from becoming wet. If one of your boots feels heavier than the other, the insulation is probably wet and they should be turned in for new ones.

Insulated boots must be tested annually to ensure they are waterproof and have less than 0.01 percent moisture in the insulation. The air valve on the outside of the boot is used to equalize pressure when worn at high altitudes and adjusted again upon return to low altitudes. To equalize the pressure in the boot, open the valve for a few seconds and immediately close it. The valve should be kept closed at all other times to prevent moisture from entering. Never open the valve unless the boot pressure around your foot causes discomfort.

You can wear overshoes over the leather boots on wet or muddy terrain. The overshoes are not as warm and dry as the insulated boots. Make sure your socks are dry and that your boots are loosely laced so as not to restrict circulation.

## 2-4. EXTENDED COLD-WEATHER CLOTHING SYSTEM (ECWCS) (TEST)

The ECWCS is issued as a one-for-one replacement for the current standard cold-wet clothing system. System components use the most recent materials and technology. The result is a reduction in weight and bulk from the current cold-wet clothing system without loss of protection from the cold. The ECWCS uses the layer approach that draws moisture away from the body and should be worn as discussed below.

### FIRST LAYER

This consists of polypropylene undershirt and shorts. They are worn next to the skin. They are designed to draw moisture away from the skin, thereby reducing conductive heat loss. You must make sure that no other item is placed between these items and your skin.

### SECOND LAYER

The second layer contains a synthetic fiber pile (SFP) shirt and polyester batting trouser liners. The SFP shirt and trouser liners provide insulation. The shirt is designed to dry quickly if it becomes wet and to provide a high degree of insulation. It should not be worn as an exterior garment in wet or windy conditions, since the loose weave provides little protection against wind chill. If the shirt gets wet, it will reduce the insulation value.

The trouser liners are designed to be worn during little or no activity. They should not be worn during heavy activity such as skiing or snowshoeing.

### THIRD LAYER

This consists of polyester batting field jacket liner and field trousers. The jacket liner should only be used for extra insulation during periods of little activity. This liner is designed to be used as a separate jacket; it should not be attached to the parka. The trousers provide a durable wind-resistant, water-repellent layer to protect the insulating underpants and trouser liners.

### FOURTH LAYER

Windproof, waterproof parka and trousers make up this layer. The parka and trousers provide windproof and waterproof shell garments to protect the first three layers.

The parka should be the outer garment, which is always worn when doing any activity outside a shelter. It is waterproof, yet the fabric is breathable so that body moisture can escape. The hood is designed to give full protection from wind and snow and does not prevent your seeing the area around you. It has underarm zippers to allow air to circulate during heavy activity. The trousers are made of the same material as the parka and are very warm. They should be worn only when necessary to prevent the field trousers from becoming wet.

### FIFTH LAYER

This consists of standard camouflage overwhites.

## 2-5. TIPS ON WEAR AND MAINTENANCE OF CLOTHING

The previous discussion on the layers of cold-wet and cold-dry uniforms describes how each system functions. In some units, the cold-dry field trouser liners are often substituted for the wool field trousers of the cold-wet system and function well in a cold-wet environment. Your commander and unit standing operating procedure (SOP) will determine what uniform and components will be worn and carried in the field.

### DRYING WET CLOTHING

Hang items separately from drying lines inside the tent. Do not place anything too close to a stove or over steaming pots; steam will cause further dampness.

### DRYING DAMP CLOTHING

Damp items can be hung on the rucksack during movement. Pin damp socks to load-carrying equipment or shirt during movement or just inside the last layer of clothing. Do not place damp clothing in sleeping bags. This will only make sleeping bags wet, and they are harder to dry than clothing.

Small items can be placed between the sleeping bag and the sleeping mat under your legs. When washing clothes without soap, dry-rub and air clothing and sleeping bags.

Never warm your feet next to an open fire. The boots will melt and burn because of their insulation before the feet become warm.

Try never to wear VB boots in a sleeping bag. Sometimes the tactical situation is such that boots must be worn while sleeping. If so, the feet must be dried and massaged with foot powder, and dry socks must be put on. Wearing the VB boot while sleeping does not let your feet dry out. After VB boots have dried, place them underneath your sleeping bag on top of your sleep pad when temperatures are below 0°F.

## 2-6. COLD-WEATHER EQUIPMENT

The sleeping equipment consists of a sleeping bag, an insulated sleeping pad, and a waterproof bag. There are two types of sleeping bags:

(1) Sleeping bag type I, intermediate cold, temperature range +45°F to +10°F, consists of a mummy-shaped bag with hood, insulated sleeping pad, and a waterproof bag.

(2) Sleeping bag type II, extreme cold, temperature range of +10°F to -50°F, is also available.

Whatever clothing is necessary must be worn with both bags to stay warm and comfortable at the lower temperature ranges. As a minimum, long underwear, sleep shirt, dry socks, and head protection should be worn while sleeping.

The insulated sleeping pad has replaced the pneumatic mattress. Use the insulated sleeping pad under the sleeping bag for warmth and comfort and to keep your sleeping bag dry.

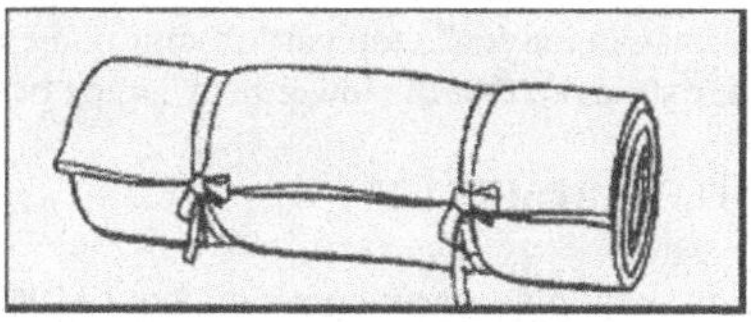

**Figure 2-4. Sleeping pad.**

The bag must be kept dry if it is to keep you warm. Sleeping bags become wet from outside moisture, from sweating on the inside of the bag, and from water escaping from the mouth or nose into the sleeping bag.

To keep outside moisture from wetting the bag, place protective insulation, pine or spruce boughs, a poncho, or an insulated sleeping pad under the bag. Avoid sweating by wearing the least amount of clothing to keep warm inside the bag, and by using the proper sleeping bag for the temperature conditions. Do not breathe into the bag or moisture will collect and wet the bag or form ice crystals. If your face gets cold, cover it with a towel, muffler, balaclava, or sleeping hood. The sleeping shirt with hood is specifically designed for use with the types I and II sleeping bags.

When it is not in use or being carried, open the sleeping bag wide so that fresh air can get into it. Turn it inside out so that the sun and wind can dry it. Always shake the bag out and fluff it up before and after use to provide for more trapped air between the layers of the wind-resistant/water-resistant outer shell.

Always carry your sleeping bag in the waterproof carrying bag. This protects the bag and helps keep it dry. Ensure that the top bag is securely tied to prevent snow or other moisture from getting inside and wetting the sleeping bag. The following suggestions will help you keep warm and comfortable in the sleeping bag:

- Eat just before getting into the bag. This gives you energy and helps you to stay warm.
- Always relieve yourself just before going to bed. The act of getting out of a warm sleeping bag to go outside can be a chilling experience.

### PACKING YOUR RUCKSACK

When you load your rucksack for skiing or climbing, pack heavy items at the bottom and next to the frame. This places the weight on your hips, which is necessary for good balance. Place hard or sharp objects inside the load where they will not rub on the bag and your back. When walking or snowshoeing, raise the center of gravity and take more of the load on your shoulders. Articles that are often needed are placed in the outside pockets for easy reach. Keep maps and other flat objects in the flap pocket.

### ADJUSTMENT

Adjust the straps so that you can move the top part of your body and swing your arms freely. Adjust the shoulder straps so that the lower back straps fit just above the belt line.

## 2-7. LOAD-CARRYING EQUIPMENT

The all-purpose lightweight individual carrying equipment (ALICE) rucksack comes in sizes medium and large. Make sure you are equipped with the large ALICE in a cold environment.

In cold-weather operations, you should follow the packing list set forth in your unit SOP to organize your rucksack. The main consideration is to avoid having to unpack during halts to get frequently needed items. The following illustrates a method of organizing your rucksack:

- **Small external pockets.** These contain small high energy foods (candy, cereal bars) to be eaten while on the move.
- **Large external pockets.** These contain rations to be eaten for morning and evening meals. Take extra socks, scarves, and a spare cap. The openings behind the pockets can be used for skis or other items.
- **External attachment points.** The sleeping pad is attached to the bottom of the pack or under the top flap (wrapped inside sandbags) and other combat items as required.
- **Top flap.** It contains camouflage overwhites and pack cover.
- **Main compartment.** Place the sleeping bag in the bottom of the rucksack. Spare clothes are placed in the upper half of the compartment for easy reach.
- **Camouflage.** A white cotton cover is provided that covers the entire rucksack, excluding the shoulder straps. Use it when wearing your overwhites.
- **Care and maintenance.** Check your rucksack for damages before and after an operation. Repair damages immediately or turn in through supply channels for a replacement.
- **Emergency quick-release of the rucksack.** When quick removal of the rucksack is needed, unsnap the fastener. Pull down on the quick-release fastener on the left shoulder strap, shift the rucksack toward the right shoulder, and let it drop to the ground.

- **Cargo support shelf.** The rucksack can be converted to a cargo carrier by removing the combat pack and replacing it with the cargo support shelf. You can carry square, rectangular, or box-type items such as radios, water cans, and fuel cans.

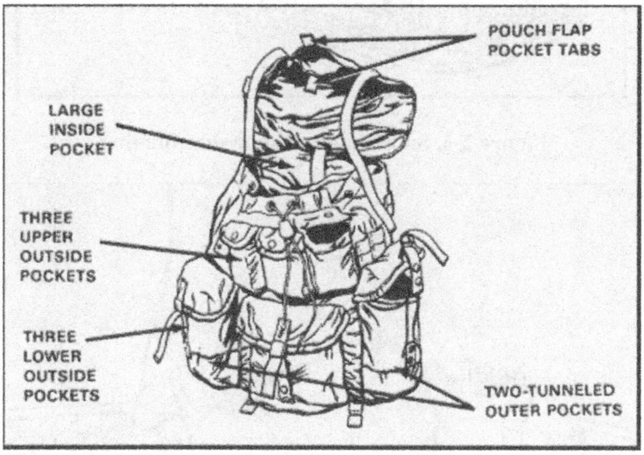


**Figure 2-5. Load-carrying equipment.**

## 2-8. OVER-THE-SNOW MOVEMENT EQUIPMENT

Before you are instructed to use snowshoes or skis, your chain command will first determine if movement by foot is possible. In shallow snow (8 inches or less) or heavily crusted snow, movement on foot for inexperienced skiers provides better maneuverability with the least effort.

### SNOWSHOES

Snowshoes consist of a white magnesium frame laced with nylon-coated steel cable. They are about 48 inches by 12 inches weigh 3.3 pounds with bindings. When properly adjusted, the toe of the boot will be about halfway across the opening in the snowshoe; the toe strap should be firm, but not tight enough to restrict circulation; and the heel straps should permit some side-to-side movement. When used correctly, snowshoes provide better mobility than walking. In most instances, skis are preferred over snowshoes. When using snowshoes, you cannot back up or swim; be careful when moving over thin ice. The traction needed to ascend or descend with snowshoes can be increased by waxing. The use of ski poles with snowshoes aids movement and reduces fatigue.

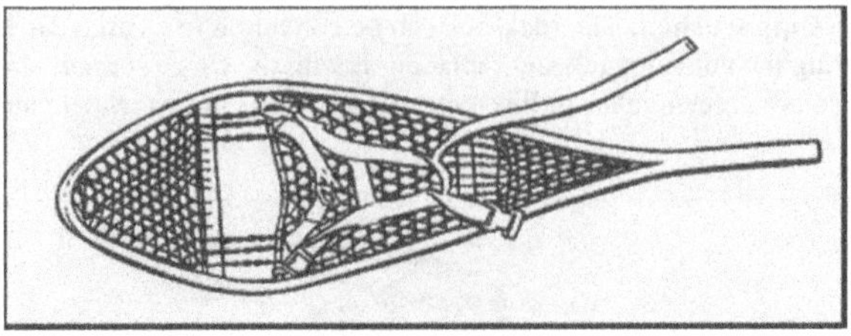

**Figure 2-6. Snowshoe and snowshoe binding.**

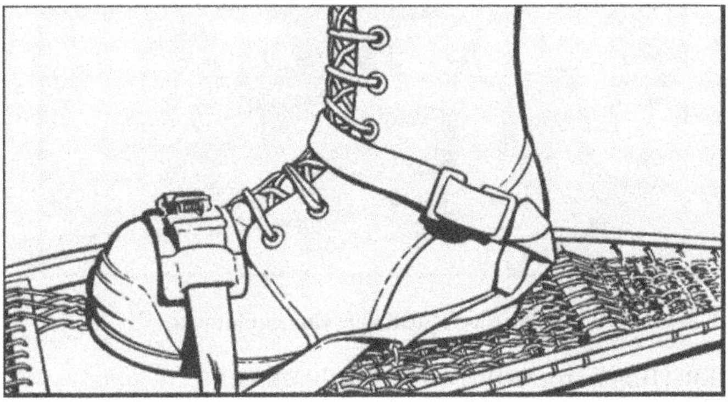

**Figure 2-7. Snowshoe binding attached to VB boot.**

## SKIS

The standard military skis are about 208 centimeters (82 inches) in length, and have a hole in the tip for towing, a groove in the heel for attaching mohair climbers, and metal edges.

## SKI POLES

Ski poles can be very useful in steep terrain. The military ski pole consists of a steel shaft with rubber handgrips and baskets, and a leather wrist strap. The poles come in three lengths: 130 centimeters (51 inches), 137 centimeters (54 inches), and 147 centimeters (58 inches). When fitting the poles, you should stand on the floor and place the pole at your side; it should come to the point of your shoulder (about 1 inch below the top of the shoulder).

### MOHAIR SKI CLIMBERS

The mohair climbers are attached to the bottom of the skis for added grip when climbing steep slopes. The climbers consist of a cotton-backed mohair pile and are attached to the skis by five web straps.

## 2-9. MISCELLANEOUS EQUIPMENT

### SUNGLASSES

Polarized sunglasses should always be worn on bright days when the ground is covered by snow. Failure to wear the sunglasses, even on overcast days when the need seems less, may result in blindness.

### PROTECTIVE MASK

There is an M4 cold-weather winterization kit available for your protective mask. Make sure you install this winterization kit for operations in the northern environment.

### CANTEENS

In cold weather the plastic canteen will freeze very quickly if carried in the standard fabric carrier. Since the canteen cannot be placed directly on heat, there is no quick method for thawing. Whenever possible, the canteen should be carried in one of the interior uniform pockets, or wrapped in clothing and placed in the rucksack to ride against your back. In extreme cold, do not fill the canteen over two-thirds full; this allows for expansion if the water freezes.

The insulated canteen is a one-quart capacity stainless steel vacuum bottle. It is provided with a cotton duck cover and a stainless steel cup. A new arctic canteen will keep water from freezing at -40°F for 6 hours. The following are some tips in using the arctic canteen:

- Fill the canteen using your canteen cup.
- Fill to within half an inch of the neck.
- Do not drop or dent.
- Put it in your sleeping bag at night.
- Carry the canteen in an M60 bandoleer near your body.

**Figure 2-8. Insulated canteen with cup.**

# CHAPTER 3

# Tents and Heating Equipment

***"A well-sheltered defender enjoyed a marked advantage over his exposed attacker..."***

***Ironside, 1918***

## 3-1. GENERAL

To conduct successful military operations in cold weather and maintain a high level of combat efficiency, you must have heated shelters. Tents and stoves are a vital part of cold-weather equipment.

## 3-2. TENT GROUP EQUIPMENT

The following is a list of the tent or ahkio group for an infantry squad:

1 Ten-man tent with liner
1 Yukon stove
1 Metal or plywood stove base
2 Five-gallon gasoline cans (one with white gasoline per platoon)
1 Five-gallon water can (filled)
2 Cases, meals, ready-to-eat
1 Climbing rope (120 feet)
1 Box of candles
2 Squad cook sets
2 One-burner squad stoves
2 Axes
2 Saws
1 Hatchet with sheath
1 D-handle shovel
1 Sledgehammer
1 Pick
1 Tripod for gas can (set)
3 Pairs of snowshoe bindings (spare)
1 Lantern (extra mantles)
1 Box of soda bicarbonate (fire extinguisher)
1 Roll of soft wire
5 Plastic bags
3 Rolls of toilet paper

## 3-3. AHKIO

The ahkio is the squad's method of transporting the tent group equipment in the winter. It is a 200-pound capacity fiberglass sled that can be used for transporting tent group equipment, for carrying weapons and ammunition, as a firing platform, and for the evacuation of casualties.

## 3-4. PACKING

When packing the ahkio, the load must be evenly distributed. Place heavy equipment on the bottom and slightly to the rear of center. Loading lighter equipment toward the top will prevent the ahkio from becoming top heavy. Tools such as shovels, axes, and saws are packed at the sides for easy access in trailbreaking; however, caution must be taken to ensure that these items do not stick out. The canvas cover is folded over the load. To keep the load from shifting and to prevent snow from getting under the cover, lash the load tightly by crisscrossing the lashing rope from the lashing rings on one side of the sled to the rings on the other side. An additional ahkio, issued on a basis of one per two 10-man tents, should be used to carry camouflage nets and support systems. In very wet or dry snow, the movement of an ahkio is easier when wax is applied to the running surface.

## 3-5. TEN-MAN ARCTIC TENT

The 10-man arctic tent with liner is a six-sided pryamidal tent supported by a telescopic pole. The tent is designed to accommodate 10 men and their clothing and equipment. It will hold additional personnel by leaving packs and equipment outside the tent, or by lowering the center pole, which allows the sidewalls to spread out to cover more ground surface. The tent has two doors, which permits joining of tents for access from one tent to another. A snow cloth is attached to the bottom of the sidewalls for sealing the tent to the ground. The snow cloth should always be to the outside of the tent with pine or spruce boughs underneath, or it may freeze to the ground. This allows easy escape in case of an emergency if the doors are blocked. The tent is ventilated by four built-in ventilators on opposite sides and near the top of the tent.

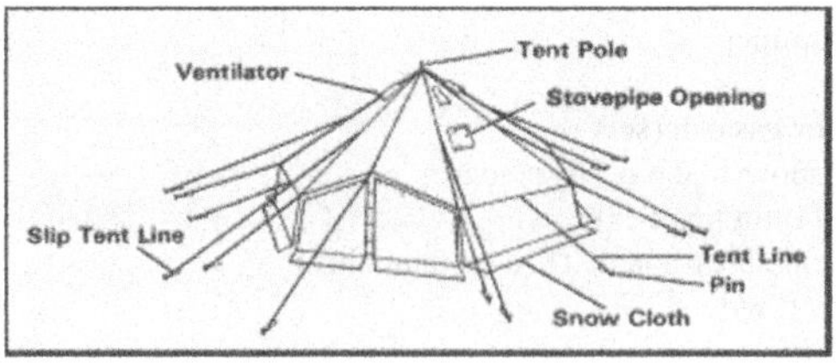


**Figure 3-1. Ten-man arctic tent.**

## 3-6. BUILDING ARCTIC TENTS

The shelter is easily built by four trained soldiers. Consider the following when pitching or striking tents in snow or on frozen ground.

Snow should be cleared to the ground surface to obtain a lower silhouette and gain advantage of ground temperatures, which are generally warmer than upper air temperatures. Enough pine or spruce boughs should be placed on the tent and floor, and under the snow covers, for insulation and comfort. If an insulating material is not used, the heat of the stove will thaw the ground insulation in a sea of mud. The snow covers will freeze to the ground, preventing escape from the tent in case of an emergency. When it is impractical to remove the snow to the ground level, an adequate tent site can be prepared by packing the snow with skis or snowshoes until a firm base is provided for pitching. The center pole is placed on a log or other suitable support to keep it from sinking into the snow.

In open terrain, with a strong wind, it will be necessary to build a 1-meter-high snow wall on the windward side of the tent to protect it from the wind. The tent is pitched with the entrance angled about 45 degrees from the downwind side. When the tent is set up, the snow cloth around the bottom of the tent should be spread out so that snow or logs can be placed on it to help in anchoring the tent.

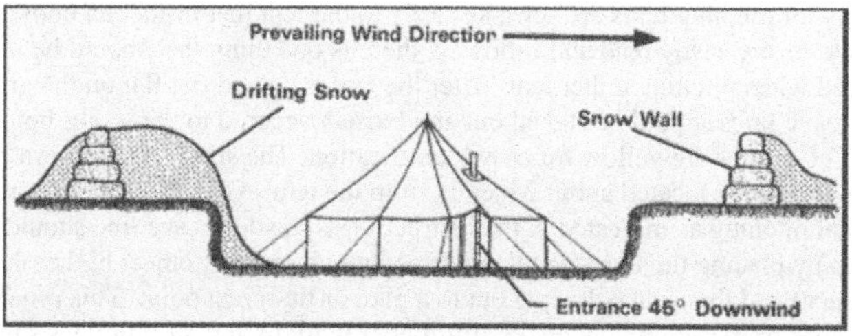


**Figure 3-2. Windbreak.**

Driving tent pins into frozen ground is impossible. Tents may be pitched and anchored by attaching the tent lines to trees, branches, logs, or stumps. If these are not available, holes are dug into the snow for the purpose of using "deadmen." This is done by digging a hole into the snow large enough to insert a pole or log about 1 meter long with the tent line attached. The hole is filled with snow and well packed. The packed snow freezes, securely anchoring the tent. In powder or dry snow, water would be poured into the hole.

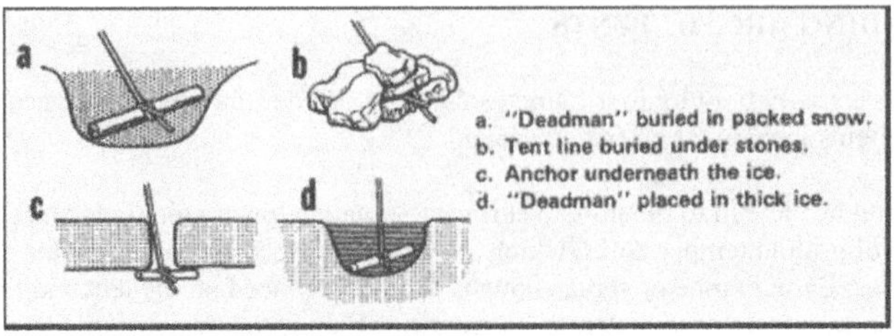


**Figure 3-3. Deadmen.**

The smaller tent (five-man) provides the commander more flexibility in distribution of his forces. You can never be assigned a tactical mission in the arctic without provisions for a shelter. This tent enables a smaller unit to be assigned an independent mission, such as patrol or reconnaissance mission. It is also six-sided, pyramidal, and supported by a telescopic tent pole. It is designed to accommodate four to five men and their clothing and equipment. The tent can provide shelter for a rifle squad or other similar unit when rucksacks are placed outside the tent.

## 3-7. INSTRUCTIONS FOR PITCHING TENTS

Instructions for pitching tents are sewn securely to the tent just inside the door. Although the instructions are easily read and followed, there is one thing that should be remembered when pitching either tent. After the tent is spread out flat on the ground, the six corner eave lines should be staked out and loosely secured to the stakes before the tent is raised. These lines are yellow for easy identification. The stakes or tie-down points for these lines should be located about 2 meters from the tent. After this is done, proceed with the tent pitching as indicated in the instructions. The door eave line should be propped up by placing the line over a pole (tree branch or other object higher than the door entrance) and then stake the line out to a pike or tie-down point. This procedure keeps the door from sagging and makes the zipper work better. Drying lines should be attached to the liner inside the tent to allow drying of clothes.

In winter the tent will normally be covered with ice and snow, which must be removed before the tent is struck. This can be done by shaking the tent or by beating it with a mitten or stick. If the snow cloth is frozen to the ground, the snow and ice around it must be removed by chopping or shoveling. This must be done carefully to avoid damage to the snow cloth. One way to do this is to ease the shovel between the cloth and the ground and gently pry the cloth away from the ice.

When the bottom of the tent is properly sealed and the door is zipped, moisture will form on the inside of the tent and collect on clothing and equipment. Carbon monoxide and fumes from the stove can accumulate to a dangerous degree. To avoid this, ensure that the ventilators near the peak of each tent are open.

The roof of the shelter must be steep enough to shed snowfall. With both the 5-man and 10-man tent, the area that will become the floor of the tent is dug, packed, and insulated with pine, spruce boughs, or similar material. Snow walls should then be built to protect against wind.

**3-8. YUKON STOVE**

The Yukon stove is used to heat the 10-man, 5-man, and GP small tent. In addition to providing heat, the top surface of the stove and the area beneath the stove may be used to heat rations and water. The stoves use standard MOGAS but may also be operated with diesel fuel or JP-4 without modification. NEVER MIX FUELS. When solid fuels (wood, coal, etc.) are used, the stove must be modified by removing the burner unit, closing the burner opening, and turning over the wire grate so there is space below for draft and ashes.

**WARNING**

**DO NOT USE A SPARK ARRESTER ON THE M1950 YUKON STOVE.**

A flue cap, issued as a component part of the Yukon stove, shields the stovepipe top from the wind and prevents the wind from blowing smoke or gases down the pipe into the stove and tent. The flue cap is equipped with three 4 1/2-meter lines that are used to steady the stovepipe and anchor it in high winds. These guy lines must be anchored to the tent or tent ropes, not to the ground or nearby trees. The stovepipe hole cover must be rolled up and tied off to prevent nylon ties from touching the stovepipe.

A firm base for the stove must be provided to keep it from melting down into the snow. The base should be 35 centimeters by 70 centimeters and be made of lightweight metal or salvaged plywood. If plywood is used, it should be covered with metal or aluminum foil. If those materials are not available, a base can be made out of small logs about 10 centimeters in diameter.

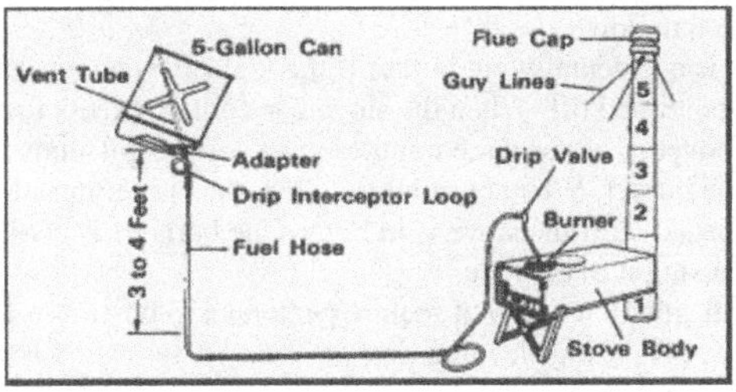

**Figure 3-4. Yukon stove.**

A simple stand for holding the fuel can is built by obtaining three poles 2 meters long. The poles are tied about two thirds of the way up and then spread out to form a tripod. The fuel can should be at least 1 meter higher than the stove. If the fuel can is wobbly or if there is a wind, the can should be tied to this tripod. Make sure the can is tilted so that the air is trapped in the uppermost corner.

**Figure 3-5. Tripod for gas can.**

## 3-9. PRECAUTIONS

The following precautions must be observed when the Yukon stove is being used:

- All stovepipe connections must be tight, and the tent stovepipe shield properly adjusted. The stove must be level to ensure that the burner assembly will spread the flame evenly throughout the stove.
- Ensure that the fuel line is made of natural rubber and suitable for extreme-cold operation. It must be protected so it cannot be accidentally pulled loose and must have a drip intercept loop to prevent fuel from running down the line and into the tent.
- Rate of fuel flow must be checked regularly. It will change with the fuel level and will need adjustment. All fuel supplies must be kept outside the tent.
- The stove should never be left unattended when it is burning. Allowing the stove to become too hot may cause the stove body to become overheated and warped and burn the tent down.
- If the flame is accidentally put out, or if the fuel can is being changed, the drip valve must be turned off. When the stove has cooled, excess fuel inside the stove must be removed. Two or three minutes must pass so gas fumes escape before relighting the burner. When burning gasoline, the burner must be cool before relighting the stove. If the stove is lit before the burner has cooled, the fuel will vaporize, causing it to explode.
- If the tactical situation does not require personnel to be awake 24 hours a day, turn the stove off during sleeping time.

The following are precautions when **burning wood or coal**:

- Fuel should be fed in small amounts until there is a hot bed of coals.
- The stove should not become overheated.
- Gasoline should not be poured on the wood or coal inside the stove to hurry starting the fire.
- Ashes should be removed regularly and not allowed to accumulate higher than the bottom of the grate.
- NEVER MIX FUELS.

## 3-10. SQUAD STOVE, M1950

The one-burner squad stove is a cooking and heating unit for two to five men. It is used when operating in an isolated or forward area when separated from the ahkio group. The stove is small, compact, and lightweight, and it will operate on either MOGAS or white gasoline. The initial pressure for its operation is generated by a few strokes on the hand pump. Thereafter, the pressure is generated by heat from the burner evaporating the fuel in the stove tank.

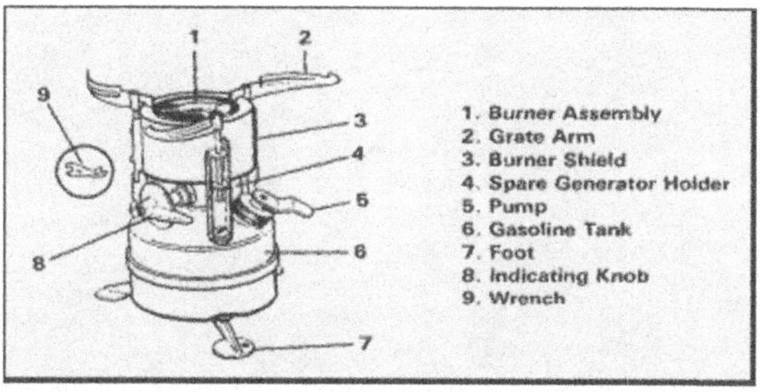


**Figure 3-6. Gasoline cooking stove, M1950.**

To operate the squad stove, do the following:

- Snap out the three feet and set stove on a level surface.
- Raise the grate arms and turn them outward so that the inner tabs can be engaged in the slots in the burner shield.
- Close the valve by turning the indicating knob clockwise to the OFF position.
- Unscrew the pump caps and remove the pump assembly.
- Fill the gasoline tank three-fourths full.
- Wipe off any gasoline that may have been spilled during filling and replace the pump cap.
- Pump 10 strokes to build up initial pressure.
- Turn the indicating knob to the ON position for 3 seconds. This allows the gasoline to rise through the generator to the burner and preheated cup.

- Turn the indicating knob to the OFF position to prevent excess fuel from accumulating in the burner assembly.
- Light the wick and allow the flame to burn for about 3 minutes or until the flame turns blue.
- Turn the indicating knob to the LIGHT position. Have a second match ready to relight the burner immediately if the flame goes out.
- When the flame in the burner turns blue, turn the indicating knob to the ON position.
- Pump an additional 10 strokes to maintain pressure within the tank. The stove is now ready for use.
- If the flame burns unevenly or dies during operation, turn the valve to the OFF and ON position several times, relight quickly, and pump several additional strokes.
- To turn the stove off, turn the indicating knob to the OFF position and allow the flame to go out.

CHAPTER 4

# Small Arms and Ammunition

***"Rifles and machine guns and even the breechblocks of artillery became absolutely rigid ... and steel parts cracked."***

***General Rendalic, Eastern Front, 1942***

## 4-1. GENERAL

Cold weather will greatly affect the operation of your weapon. You must be aware of these effects to ensure employment of your weapon under poor weather conditions.

## 4-2. CARE, CLEANING, AND MAINTENANCE

Your weapon will work properly under extreme cold conditions if you take care of it. Normal temperate zone lubricants thicken in cold weather, and stoppages or sluggish actions of your weapon will result. During the winter, disassemble your weapon completely, clean it thoroughly, and remove all lubricants. Oil the entire weapon with a light coating of cleaner lubricant preservative (CLP).

| CAUTION |
|---|
| **DO NOT USE CLP ON A MORTAR OR ANY COMPONENT PART OF A MORTAR.** |

**Figure 4-1. Weapons will 'sweat" when taken in to a heated shelter.**

A main problem is to keep snow and ice, which may disable your weapon, out of the working parts, sights, and barrel. When the weapon is not in use, keep it out of the snow either in an improvised weapon stand or in a tripod arrangement. If this is not possible, stand it up with the butt end in the snow.

Condensation ("sweating") will form on the weapon when you take it from the cold into a heated shelter. The condensation will continue for about 1 hour, after which you should clean the weapon. If you don't, the condensation will freeze when you take the weapon outside, and it may not fire. The weapon must be kept as near the floor as possible to lessen condensation. If parts of the weapon do become frozen, warm them and move the working parts slowly until they are unfrozen. If you can't warm your weapon, remove all ice and snow and move the parts gently until action is restored. The use of CLP will prevent malfunctions by eliminating condensation.

When firing, do not let the hot parts of your weapon come in contact with the snow. The snow will melt and, as the weapon cools, will form ice. When changing barrels, don't lay them on the snow.

To prevent bipods and tripods from sinking into the snow, improvise for added flotation. This is done by placing them on snowshoes, ski poles, branches, ahkios, etc. Baseplates should be on solid ground or sandbags to prevent sinking into the snow and damage to the weapon.

**Figure 4-2. Bipods and tripods will sink in the snow.**

## AMMUNITION

Small arms ammunition should be kept at the same temperature as your weapon. Ammunition clips and magazines must be cleaned of all oil and preservative and checked often. All ice, snow, and condensation must be removed. Ammunition boxes must be kept closed in order to keep out snow and ice.

Stored ammunition should be kept in packed boxes, raised off the surface by placing them on boughs, cardboard, or other similar material, and covered to protect them from the snow. Ammunition should be well marked for easy location and identification in case it becomes covered with snow.

Ammunition resupply may become restricted. Everyone must be aware of the necessity for ammunition economy and fire discipline. Loaded clips, magazines, or single rounds dropped in the snow become quickly lost; therefore, careful handling of ammunition is essential. Never waste your ammunition. Make every round count.

When contact with the enemy is imminent, troop compartments and aircraft should be kept as close to freezing as possible to avoid condensation. However, many munitions have limitations at lower temperatures, which must be recognized in training.

# CHAPTER 5

# Rations and Diet

***"I ate my boots and a pair of pants ... tea and seal-skin gloves for dinner."***

***Greely, Arctic Expedition, 1883***

## 5-1. GENERAL

Most of what we eat and drink is used in maintaining our body heat, while only a small portion is used in producing energy for physical work. You must ensure adequate caloric intake in cold-weather operations. About 4,000 calories per day are necessary for personnel performing physically demanding work in the cold. Efficiency may drop rapidly if this level is not maintained.

The body loses liquid at a very fast rate in arctic conditions, regardless of how carefully you adjust and ventilate your clothing. The exertion of movement on foot, preparation of bivouac sites and defenses in the snow, etc., take a toll in sweat and loss of moisture in the breath. These liquids must be regularly replaced, preferably by hot drinks, which provide extra calories if they contain sugar.

## 5-2. RATIONS

Rations provide the needed calories to live and fight effectively. When eaten in their entirety, rations contain the right amount of carbohydrates, fats, protein, and vitamins. The proper intake of these essential items depends on the entire ration being eaten in properly spaced meals. The lack of concern regarding eating caused by the cold, combined with the difficulties and inconvenience of cooking, may tempt soldiers to miss meals. The principles of sound leadership and discipline in cold weather require that meals be prepared and that the entire ration be eaten, warmed when possible. Save snacks (cookies, chocolate bars, etc.) for between meals and when on the march.

## 5-3. LIQUIDS

Because there is a large amount of fluid loss in the cold, you should drink a minimum of 3.5 quarts of water a day when heavy physical activity is involved. Water is usually available either from streams or lakes or by melting snow or ice. A limiting factor may be fuel needed to melt the snow or ice. To save time and fuel, water should be obtained from running streams or a lake. The milky water of glacial streams must be allowed to stand until the sediment settles. When a hole is cut into the ice to get water, it should be covered by a snow block or something similar to slow refreezing.

Ensure that as much of the daily liquid intake as possible is hot drinks, such as soups and cocoa. Your local command may request a special food allowance of warming and cooling beverages. Main meals should begin with soup, and between-meal snacks should include a hot drink.

When running water is not available, ice or snow must be melted. Ice produces more water in less time than snow. When melting snow, first put a small amount into the cooking pot; then more can be added. Continue this process until there is enough water. It must be purified by boiling rapidly for 15 minutes.

When heating water, use all available stoves, since this is a most time-consuming operation. Melting and boiling enough snow for a drink may take 30 to 40 minutes. Areas designated for ice or snow that is to be used for water must be sited far upwind of the latrine and garbage disposal sites. Before going to bed, ensure that stoves are filled and enough water has been prepared for breakfast. Do not let other soldiers eat snow or ice; doing so lowers the body's core temperature.

Never drink alcohol on operations or when exposed to low temperatures. The aftereffects of alcohol can lead to a feeling of false security. You may forget the main rules to prevent cold injury, like wearing your gloves or not going outside without shoes to relieve yourself.

# CHAPTER 6

# Hygiene and First Aid

***"...the first frostbite cases have made their appearance. We amputated two feet and will amputate many more."***

***Amosoff***

## 6-1. GENERAL

The care of your body needs special attention in cold weather. Ensuring cleanliness in the field, particularly in the cold, is important. It is easy to let yourself go for long periods without bathing, but if you wish to be part of an effective combat force, you must keep clean.

## 6-2. HYGIENE TIPS FOR THE SOLDIER

Face, hands, armpits, and crotch should be washed daily. If water is not available, take a "bath" using cornstarch or rub your body briskly with a dry towel. The cornstarch removes excess oil and perspiration from skin and hair.

Shave regularly -- daily if water is available or every 2 to 3 days if water is in short supply. Shaving just before sleeping will allow the face maximum recovery time before going out into the cold, reducing the possibility of superficial frostbite.

Teeth should be cleaned daily. If a toothbrush is not available, a clean piece of cloth wrapped around a finger or a chewed twig may be used as a substitute.

Underwear should be changed at least twice a week. If it is not possible to wash your underwear, crumple it, shake it, and air it for at least 2 hours.

Ensure that all eating and cooking utensils are cleaned after each use.

In the dry arctic you will find the need to oil your unprotected and exposed skin once every week (with an oil such as suntan oil).

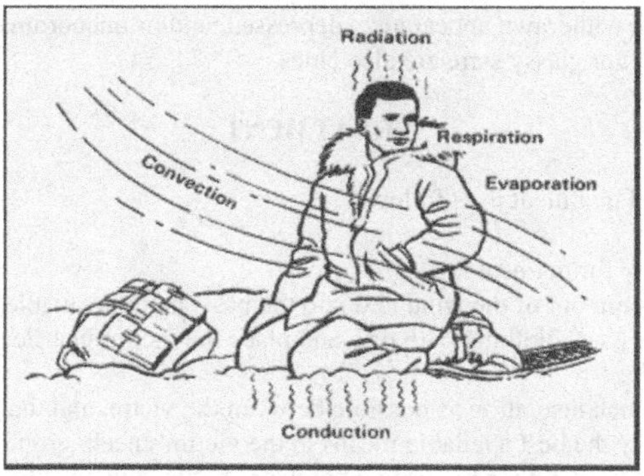


**Figure 6-1. Mechanisms of heat loss from the body.**

## 6-3. HYPOTHERMIA

Hypothermia is a lowering of the temperature of the body's inner core. This happens when the body loses heat faster than it can produce it. You must be able to recognize the symptoms of hypothermia and to treat the condition immediately.

### PREVENTION

To prevent hypothermia, adhere to the following:

- Stay physically fit.
- Keep active.
- Use the uniform properly and keep it dry.
- Eat properly and often.
- Drink plenty of liquids, at least 3.5 quarts per day, when performing physically demanding work.
- Be prepared for and know how to deal with quick changes in the weather.
- Bivouac early before judgment is weakened by fatigue.

### SYMPTOMS

Be aware of the following symptoms:

- Shallow breathing or absence of breathing.
- Faint or undetectable pulse.
- Victim is first cold, then stops shivering.
- Loss of feeling.
- Poor coordination; victim has trouble walking.
- Impaired speech.

- Confusion, withdrawn appearance, depressed, and/or uncoordinated. An uncaring attitude and/or glassy stare are also clues.

**TREATMENT**

The recommended treatment is as follows:

- Prevent any further heat loss.
- Get the victim out of the wind and into the best shelter available.
- Replace any wet clothing with dry, and place the victim in a sleeping bag, if one is available.
- Place as much insulation as possible between the victim and the ground.
- Add heat by the best available means to the victim's neck, groin, and sides of chest. This can be done by using a hot water bottle, Yukon stove, campfire, or your own body heat.
- Provide the victim with something warm and nutritious to drink if he is conscious. Calories may be added by using sugar in hot, sweet drinks.
- DO NOT massage the victim.
- DO NOT give alcohol to the victim.
- Evacuate the victim to the nearest medical treatment facility as soon as possible. This condition is life threatening until normal body temperature is restored.

## 6-4. FROSTBITE

Frostbite results when tissues freeze from exposure to temperatures below 32°F. The degree of injury depends upon the wind-chill factor, length of exposure time, and proper protection. Persons with a history of cold injury are prone to frostbite. There may be a tingling, stinging, an aching sensation, or a cramping pain. The skin first turns red and then becomes pale gray or waxy white. Frostbite can be classified as superficial or deep.

**WIND CHILL**

The combination of wind and low temperatures creates a condition known as wind chill. For example, with the wind calm and a temperature of -20°F there is little danger of wind chill. If the temperature is -20°F and there is a wind of 20 knots, the equivalent chill temperature is -75°F!

Under these conditions there is GREAT DANGER and your exposed skin can freeze within 30 seconds. You also create your own wind when you walk, run, or ski. As examples, walking creates a wind speed of about 5 mph and running 10 mph. Wind is also created by moving vehicles, aircraft propellers, and rotors. The wind-chill chart is provided to show how wind and temperature combine to create wind chill; use it. Cold injury can disable you like an enemy bullet.

| WIND SPEED | | COOLING POWER OF WIND EXPRESSED AS "EQUIVALENT CHILL TEMPERATURE" | | | | | | | | | | | | | | | | | | | | |
|---|---|---|---|---|---|---|---|---|---|---|---|---|---|---|---|---|---|---|---|---|---|---|
| KNOTS | MPH | TEMPERATURE (°F) | | | | | | | | | | | | | | | | | | | | |
| CALM | CALM | 40 | 35 | 30 | 25 | 20 | 15 | 10 | 5 | 0 | -5 | -10 | -15 | -20 | -25 | -30 | -35 | -40 | -45 | -50 | -55 | -60 |
| | | EQUIVALENT CHILL TEMPERATURE | | | | | | | | | | | | | | | | | | | | |
| 3-6 | 5 | 35 | 30 | 25 | 20 | 15 | 10 | 5 | 0 | -5 | -10 | -15 | -20 | -25 | -30 | -35 | -40 | -45 | -50 | -55 | -60 | -70 |
| 7-10 | 10 | 30 | 20 | 15 | 10 | 5 | 0 | -10 | -15 | -20 | -25 | -35 | -40 | -45 | -50 | -60 | -65 | -70 | -75 | -80 | -90 | -95 |
| 11-15 | 15 | 25 | 15 | 10 | 0 | -5 | -10 | -20 | -25 | -30 | -40 | -45 | -50 | -60 | -65 | -70 | -80 | -85 | -90 | -100 | -105 | -110 |
| 16-19 | 20 | 20 | 10 | 5 | 0 | -10 | -15 | -25 | -30 | -35 | -45 | -50 | -60 | -65 | -75 | -80 | -85 | -95 | -100 | -110 | -115 | -120 |
| 20-23 | 25 | 15 | 10 | 0 | -5 | -15 | -20 | -30 | -35 | -45 | -50 | -60 | -65 | -75 | -80 | -90 | -95 | -105 | -110 | -120 | -125 | -135 |
| 24-28 | 30 | 10 | 5 | 0 | -10 | -20 | -25 | -30 | -40 | -50 | -55 | -65 | -70 | -80 | -85 | -95 | -100 | -110 | -115 | -125 | -130 | -140 |
| 29-32 | 35 | 10 | 5 | -5 | -10 | -20 | -30 | -35 | -40 | -50 | -60 | -65 | -75 | -80 | -90 | -100 | -105 | -115 | -120 | -130 | -135 | -145 |
| 33-36 | 40 | 10 | 0 | -5 | -15 | -20 | -30 | -35 | -45 | -55 | -60 | -70 | -75 | -85 | -95 | -100 | -110 | -115 | -125 | -130 | -140 | -150 |
| WINDS ABOVE 40 HAVE LITTLE ADDITIONAL EFFECT | | LITTLE DANGER | | | | | INCREASING DANGER (Flesh may freeze within 1 minute) | | | | | | GREAT DANGER (Flesh may freeze within 30 secs) | | | | | | | | | |

**Figure 6-2. Wind-chill chart.**

## PREVENTION

Frostbite is preventable if the individual soldier takes care of himself. It seldom occurs in people who are maintaining enough body heat. It is most commonly associated with an overall body heat loss resulting from poor equipment, reduced food intake, dehydration (water loss), overtiring, injury, or a combination of these factors. The feet, hands, ears, and exposed areas of the face are most easily hurt by frostbite and must receive constant care. The following are rules to help prevent frostbite:

- Do not wear tight boots and socks.
- Use the "buddy system" to check exposed areas, especially when wind is present.
- Carry extra socks and mitten liners.
- Do not stay still for long periods.
- Use caution when cold and wind are combined.
- Check feet during halts.
- Ensure that you eat well and have hot drinks as often as possible.

## SYMPTOMS OF SUPERFICIAL FROSTBITE

Redness occurs, followed by powdery flaking of the skin. Affected areas of dark-skinned soldiers may appear dull and grayish. Blister formation occurs 24 to 30 hours after exposure, followed by flaking of superficial skin.

## SYMPTOMS OF DEEP FROSTBITE

Common symptoms are as follows:

- Loss of feeling occurs in the affected area, leaving it pale yellow and waxy looking.
- When the frostbitten area thaws, it is painful.
- Frozen tissue may feel solid or "wooden" to the touch.
- When exposed to inside temperatures, blisters may appear in 12 to 36 hours.

- Discoloration (red-violet) appears 1 to 5 days after the injury.
- Gangrene usually results.

### TREATMENT FOR FROSTBITE

Determine whether the frostbite is superficial or deep. If the exposure time was short, the frostbite will probably be superficial. If the exposure time was long, the frostbite will probably be deep.

The following is a recommended treatment for frostbite:

- Move the casualty to a warm and sheltered area.
- Do not rub the frostbitten area with snow or apply cold water soaks.
- Do not rewarm the affected area by massaging or exposing it to open fire.
- Rewarm the face, nose, or ears by placing your hands on the frozen area. Rewarm frostbitten hands by placing them under clothing and against the body.
- Close the clothing to prevent further loss of body heat.
- Rewarm the feet by removing the boots and socks. Place the bare feet under the clothing and against the abdomen of a buddy. Once the feet are warmed, put on dry socks and boots, if available. If the victim must wear the wet socks and boots, he should exercise his feet by wiggling his toes.
- Loosen tight clothing and remove jewelry.
- Improve circulation by exercise.
- Do not allow the victim to use alcohol or tobacco. Alcohol increases loss of body heat; tobacco causes narrowing of blood vessels in the arms and legs.
- Reassure the victim.
- Protect frozen tissue from further cold or trauma, and evacuate the casualty to the nearest medical treatment facility.

Deep frostbite is most common and harmful to the feet and less common to the hands and ears. When deep frostbite does occur, the victim should be moved to a sheltered area and evacuated to a medical treatment facility immediately. If possible, do not allow the victim to walk if his feet are frozen and avoid treating or thawing the affected area. Thawing of deep frostbite in the field increase pain and invites infection, greater damage, and gangrene. Do not try to thaw deep frostbite; let the physician do it.

## 6-5. DEHYDRATION

Dehydration can cause serious physical problems and make you more susceptible to other problems, such as frostbite and hypothermia.

### PREVENTION

The following are suggestions to prevent dehydration:

- The **minimum** daily liquid requirement for personnel performing demanding physical work in the cold is 3.5 quarts per day.
- By the time you feel thirsty you are already dehydrated. Drink whenever possible, particularly during halts.
- Ensure that canteens are full before any type of movement.
- During periods when water is scarce, do not drink coffee in large quantities. It will pass through you quickly. Water should be the first choice.
- Check your urine spots in the snow. Dark yellow or brown indicates dehydration.

### SYMPTOMS

Common symptoms are as follows:

- Red indicates a severe condition needing immediate medical evacuation to the nearest medical treatment facility.
- Lack of appetite.
- Dry mouth, tongue, and throat.
- Stomach cramps and/or vomiting.
- Headaches.

### TREATMENT

Recommended treatment is as follows:

- Keep the victim warm.
- Give him plenty of liquids.
- Make him rest.

## 6-6. CARBON MONOXIDE POISONING

Whenever a stove, fire, gasoline heater, or internal combustion engine is used indoors, there is danger of carbon monoxide poisoning. Fresh air in living and working quarters is vital. Carbon monoxide is a deadly, odorless gas.

### PREVENTION

The following are suggestions to prevent carbon monoxide poisoning:

- Use stoves and lanterns in well-ventilated areas and tents.
- Ensure that stoves and lanterns are functioning properly.
- Do not let personnel warm themselves by engine exhaust.
- Always have windows cracked in vehicles with a heater in use.

- Use a tent guard or shut the stove off when sleeping.

### SYMPTOMS

Common symptoms are as follows:

- Headache, dizziness, confusion, yawning, weariness, nausea, and ringing in the ears.
- Bright red color on lips and skin.
- Victim may become drowsy and collapse suddenly.
- If personnel are found unconscious in an enclosed shelter, carbon monoxide poisoning should be suspected.

### TREATMENT

Adhere to the following for proper treatment:

- Move the victim to open air.
- Keep the victim still and warm.
- If the victim is not breathing, administer mouth-to-mouth resuscitation.
- Administer cardiopulmonary resuscitation, if necessary.
- Immediately evacuate the victim to the nearest medical treatment facility.

## 6-7. SNOW BLINDNESS

Snow blindness is caused by ultraviolet and ultrablue rays of the sun being reflected from a snow-covered surface into the eyes. This condition can occur even in cloudy weather. In fact, it is more likely to occur in hazy, cloudy weather than in sunny weather.

**Figure 6-3. Snow blindness.**

### PREVENTION

Prevention is easy; wear your issued sunglasses. If sunglasses are not available, make slitted glasses from cardboard, thin wood, tree bark, or similar material.

### SYMPTOMS

Common symptoms are as follows:

- A scratchy, sandy feeling under your eyelids.
- Redness and watering of the eyes.
- A headache.

### TREATMENT

Adhere to the following for proper treatment:

- Blindfold the patient using a dark cloth.
- Reassure the patient.
- Evacuate the patient to a medical treatment facility.

## 6-8. SUNBURN

Sunburn is a serious hazard because of the reflective qualities of the snow, especially at higher elevations.

### PREVENTION

Require soldiers to use a sunscreen and lip balm.

### SYMPTOMS

Redness of skin with slight swelling is one symptom of sunburn. Prolonged exposure to the sun may cause pain and blistering and, in severe cases, chills, fever, and headaches.

### TREATMENT

Soothing skin creams may be helpful if the swelling is not severe.

## 6-9. TENT EYE

Tent eye is an inflammation of the eye caused by fumes from stoves and lanterns used in a poorly ventilated tent or shelter. It is prevented by proper ventilation of tents or shelters. The treatment is fresh air.

## 6-10. TRENCH FOOT/IMMERSION FOOT

Trench foot and immersion foot are injuries caused by the prolonged exposure of skin to cold or wet conditions. In extreme cases, the skin dies and amputation of the foot or leg may be necessary.

### PREVENTION

The following helps prevent trench foot or immersion foot:

- Change to dry socks at least twice daily.
- Dry and massage your feet regularly with foot powder.
- Wear VB boots and change socks regularly.
- If leather boots are worn, dry them whenever possible.

### SYMPTOMS

In early stages, the feet and toes are cold, numb, and still, and walking becomes challenging. The feet swell and become painful.

### TREATMENT

In early stages, keep the feet dry, clean, and exposed to the air. In later stages, evacuate the victim to a medical facility.

## 6-11. CONSTIPATION

At very cold temperatures, personnel tend to put off the natural urge to have a bowel movement, resulting in constipation. This can be a painful casualty producer.

### PREVENTION

Eat fresh or canned fruit and eat regularly. Drink plenty of liquids. Try to have a daily bowel movement at your normal time.

### SYMPTOMS

Some common symptoms are stomach cramps, dizziness, and headaches.

### TREATMENT

Prevention is the best treatment. If symptoms persist, get medical help.

## 6-12. DIARRHEA

Diarrhea can be treated by drinking tea leaves with water.

## 6-13. HEAT CRAMPS

Heat cramps are caused by an excessive loss of salt in the body. This loss of salt causes muscles to spasm, and they are unable to relax. After prolonged physical activity in a cold-wet environment, the signs and symptoms of heat cramps could appear in the arms, legs, and/or stomach. It is possible to experience heat cramps in cold-weather environment when undergoing extreme physical exertion.

**SYMPTOMS**

The soldier experiences muscle cramps in his arms, legs, and/or stomach. He is pale and has wet skin and is experiencing dizziness and extreme thirst.

**TREATMENT**

Have the soldier drink one fourth of a canteen of water. Give the soldier one canteen of water with one packet of salt stirred in. Have him drink the canteen of salted water over a 30-minute period. If cramping of muscles continues, evacuate the soldier to the nearest medical treatment facility.

> **NOTE**
>
> **Do not give the saltwater to a soldier who has an upset stomach. Have him drink a canten of unsalted water.**

**6-14. HEAT EXHAUSTION**

Heat exhaustion is caused by dehydration and loss of body salt from extreme physical activity in a cold-wet environment. It is possible to experience heat exhaustion in a cold-weather environment when undergoing extreme physical exertion.

**SYMPTOMS**

The soldier may feel dizzy, weak, and/or faint. The skin feels cool and moist to the touch. He may feel nauseated or have a headache.

**TREATMENT**

The following is a recommended treatment for heat exhaustion:

- Have the soldier drink one canteen of water to relieve the symptoms.
- If he complains of cramps, give him one canteen of water with one packet of salt stirred in. Have him drink the salt solution over a 30-minute period.

> **NOTE**
>
> **Do not administer salt if the soldier has an upset stomach. Have him drink a canteen of unsalted water.**

- Loosen all tight-fitting clothes.
- Elevate the feet above the level of the heart.
- If the soldier is unable to drink water due to an upset stomach or if the symptoms have not improved within 20 minutes, evacuate the soldier to the nearest medical treatment facility.

# CHAPTER 7

# Bivouac Routine

***"Had we lived, I should have a tale to tell of the hardihood, the endurance and the courage of my companions which would have stirred the hearts of every Englishman. These rough notes and our dead bodies must tell the story."***

***CPT Robert Scott***

## 7-1. GENERAL

The essential requirements for survival and mission accomplishment in the cold/arctic environment are warmth, food, and shelter. As covered in Chapter 3, you are provided with tents that can be erected quickly. Tents can be lost or destroyed and separated from the ahkio groups in the confusion of combat. When this occurs, timber and snow, which are usually available, will be used for shelters. A detailed explanation of the bivouac routine should be contained in the unit SOP.

## 7-2. BIVOUAC SITE

Bivouac sites, other than those that are permanent nature in rear areas, either are tactical or are designated to provide living accommodations within a defended position. The tactical bivouac is usually not occupied for more than one night, is established at or after last light, and needs little preparation. It is organized for a rapid and disciplined response to enemy action. The defensive position is likely to be occupied for longer periods and must be established accordingly. The layout and placement of tents and shelters must be coordinated within the defense and in accordance with the unit SOP.

## 7-3. SITE SELECTION

The selection of bivouac areas is important and needs careful consideration. The challenge of area selection varies with the tactical situation, weather conditions, and terrain. The bivouac area will be tactically located in accordance with the principles of security and defense.

There must be concealment from air and ground observation and wind. Forested areas pose few challenges in comparison to areas that are located above the tree line. In selecting areas, ensure that local camouflage materials are available for use with the snow camouflage nets. Areas that offer better bivouac sites than others are explained below.

## FORESTED AREAS

Forests provide excellent sites. Camouflage is easy and building materials, firewood, and wood for insulation are readily available. Although forests provide the best bivouac site, temperatures there will normally be colder than those in open areas.

## MARSHY GROUND

In winter, good sites are frozen swampy areas, the banks of streams, and the shores of lakes. Some swampy areas may not freeze because of warm springs or marsh gases. Deep valleys should be avoided since colder air gathers there.

## OPEN AREAS

Avoid open areas because of strong winds, drifting snow, and poor concealment. If avoidance is not possible, then pitch tents where they can be sheltered by natural windbreaks. These are normally found in depressions or behind ridges of ice on lakes. In areas where natural wind falls do not exist, build snow walls to provide wind protection. In areas with high winds, snow gathers on the lee side, making it necessary to clear the side and tops of the tents, often to prevent the weight of the drifting snow from collapsing the tent.

## MOUNTAINOUS AREAS

Mountainous terrain is characterized by strong winds, cold, and a general lack of concealment above the timberline. The wind over head creates extensive lees near cliffs and steep rock faces. Avalanches can occur on steep slopes; avoid their likely path. Signs to look for are:

- Heavy snow overlaying a crust of ice (usually the result of a freeze at the end of a period of thaw).
- New snow on steep slopes if the temperature is rising.
- Evidence of avalanches in previous years; an area cut through the trees (branches stripped off, and/or broken tree stumps at different heights) or a smooth clear strip of ground down the mountainside.

**Figure 7-1. Location of tent.**

**Figure 7-1. Location of tent (continued).**

When contact with the enemy is likely, the unit SOP will identify specific considerations when selecting the site for a tactical bivouac. For example:

- The location of the enemy.
- Enemy patrol habits.
- The air threat.
- Dominating features that identify enemy avenues of approach.
- The wind direction that can carry sounds of movement (the bivouac site should be downwind of likely enemy avenues of approach, especially when wind speeds are low).
- Availability of natural camouflage.
- Amount of moonlight and weather conditions. In good weather or bright moonlight, the bivouac site must be in an area of deep shadow (in the middle of the woods or on the northern slope of a ridge) or the tents will be silhouetted.

## 7-4. RECONNAISSANCE OF BIVOUAC SITE

The main body should be preceded by a reconnaissance party whose task is to lay out the bivouac site. They should:

- Identify the bivouac site and a false position.
- Establish a trail plan.
- Select defensive positions.
- Mark the location of tent sites.
- Identify areas for cutting boughs.
- Act as guides for the main body on its arrival.
- Provide security during occupation by the main body.

## 7-5. TRAIL PLAN

Trail plans must be established before entering the bivouac site. Normally, the incoming trail will be extended well beyond the area of the bivouac and may lead to a false position. Leading off the incoming trail establish another trail, usually at an extreme angle to entry. The trail network that leads to each subordinate unit area, and within, must follow the inside of the tree line. This prevents enemy observation of trails from the air or ground.

**7-6. OCCUPATION**

Guides from the reconnaissance party will arrange to meet the main body and identify the layout of the bivouac site. Once the unit is in position, the unit should:

- Identify a temporary area for weapons and equipment during establishment of the site.
- Identify the exact location and type of defensive positions to be built.
- Determine exact tent sites regarding defensive positions, natural shelter, and camouflage.
- Identify sleeping areas and type of shelters to be built when tents are not used.

**7-7. BOUGH-CUTTING AREAS**

Bough-cutting areas for bedding or improvised shelters should be for all personnel in the unit. These areas must be where activity is not likely — well behind the unit's defensive area or in a decoy bivouac site.

Cut boughs from the bottom portion of trees above the snow and leave the trees standing. Be sure to cut only a few boughs from each tree. A tree stripped of its limbs is easily observed from the ground and from the air.

Cut firewood near the bough-cutting area to allow the same trail to be used. Dead spruce, pine, or squaw wood makes the best firewood. Standing deadwood burns the best and hardwoods the longest. In an arctic and subarctic environment, true hardwoods are rare. Of the trees available, birch, tamarack, and spruce are the hardest woods. Burn the upper parts of dead trees during daylight because they give off lighter colored smoke. The lower parts of the trees contain resin and tar, which burn well but make more and darker smoke. In the daylight this will allow your position to be easily identified by air and ground observation.

**7-8. GARBAGE**

All garbage should be double bagged and transported to the rear or disposed of in accordance with the unit SOP. Units should never leave evidence of their presence. Carry all trash, if necessary, until it can be disposed of.

**7-9. LATRINE**

Normally there will be a central latrine within the bivouac site, if the site is not too widely spread. One latrine can usually serve the needs of a platoon-size element. It must be placed downwind of the bivouac site, but not so far that soldiers won't use it. The area identified can be as elaborate as a pit or cross-cut-type latrine that is made windproof by branches, snowblocks, and ponchos. It should be properly camouflaged. It can also be an area where plastic bags are located. Double-bag, tie, and transport all waste to the rear for disposal as identified in the unit SOP. Site urinal areas around a "dedicated tree" or the gasoline tripod, which will hide the discolored snow and will be easily spread about when leaving the bivouac site.

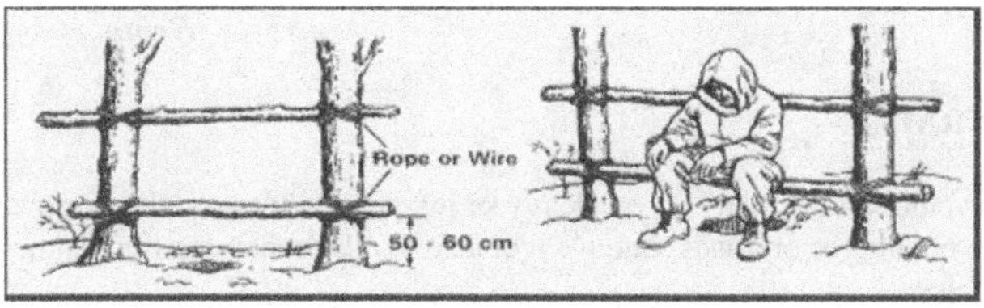


**Figure 7-2. Cross-tree type latrine.**

### 7-10. WATER POINTS AND SNOW AREAS

When obtaining water for drinking purposes by melting snow, set aside and restrict an area for this purpose only. The site must be upwind from the bivouac area and isolated from the latrine and garbage areas. The snow must be clean, white, and free of fungus from trees. NEVER use snow that is discolored. Remember — disinfect the water using purification tablets and bring it to a rolling boil for a minimum of 15 minutes.

### 7-11. STORAGE

Storage problems in the winter are increased by snow, low temperatures, thaws, and limited storage space. Space in any shelter is limited. Store only those items inside the tent that are affected by the cold, or must be immediately available. The remaining equipment should be centralized, well marked, and covered. Store ammunition and fuel separately. Place wood or metal containers on boughs or poles to keep the containers from freezing to the ground. The unit SOP will specify what equipment is placed inside or outside of the tent and where.

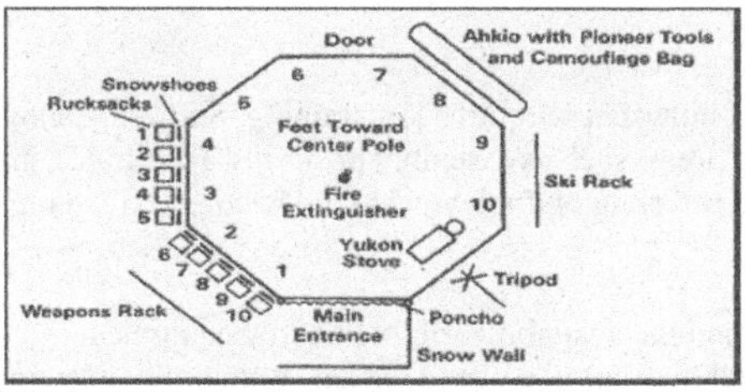


**Figure 7-3. Ten-man tent.**

# CHAPTER 8

# Individual Movement

***"Anticipate every situation which can arise in battle and think out your reaction to it well in ADVANCE."***

***Hoote***

## 8-1. GENERAL

Snow affects movement many ways. Heavy snow cover hinders or slows movement, either cross-country or on roads. Snow cover also blankets many terrain features and hides obstacles.

The effect of snow on movement varies greatly with its depth and physical characteristics. Men on foot cannot travel well over flat terrain or roads when the snow is deeper than 30 centimeters. However, movement can be improved by the use of skis or snowshoes.

| SKIS | SNOWSHOES |
|---|---|
| • Faster<br>• Less Work<br>• Harder to Learn | • Slower<br>• More Work<br>• Easy to Learn |

**Figure 8-1. Using skis or snowshoes.**

## 8-2. SKIS

Skis provide greater speed in moving, particularly over cleared trails. Troops mounted on skis and towed behind small unit support vehicles, or similar vehicles (ski-joring), are effective means for rapid cross-country movement. In heavy vegetation, skiing will always be more physically taxing than snowshoeing.

## 8-3. SNOWSHOES

Learning how to use snowshoes requires less training. Snowshoe movement is more practical in confined areas, such as assembly positions, field trains, and mortar and artillery positions. If traveling on foot, you should become skilled on both skis and snowshoes.

Your ability to use, and the availability of, over-snow equipment is a major factor in learning to travel well over snow-covered terrain. With previous training in cross-country skiing and if equipment is available, travel on skis is recommended. In most snow conditions and over most types of terrain, skis provide the fastest and most energy-saving form of travel. Use of snowshoes requires little training, but your speed will be slower and travel more exhausting.

Snowshoes are **more** exhausting than skis. However, a light crust on the surface of the snow prevents skis from sinking and provides for fast and easy skiing. A crust hard enough to support a man makes travel on foot feasible, but even then travel on skis is recommended.

In a survival situation, improvise equipment for travel if snow is loose and deep. Make snowshoes of willow or other springy wood using a wood separator and thong, wire, cord, etc.

**USE SKIS WHEN:**

- Speed is essential.
- Long distance must be traveled.
- Stealth is necessary.
- Conditions allow.

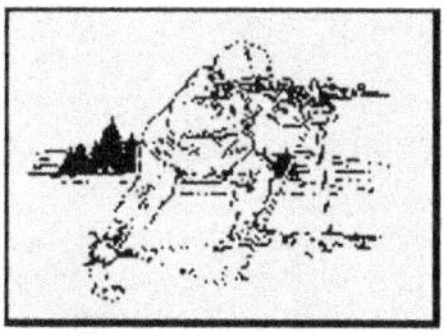

**USE SNOW SHOES WHEN:**

- Moving through heavy brush.
- Speed is not essential.
- Troops are not proficient on skis.

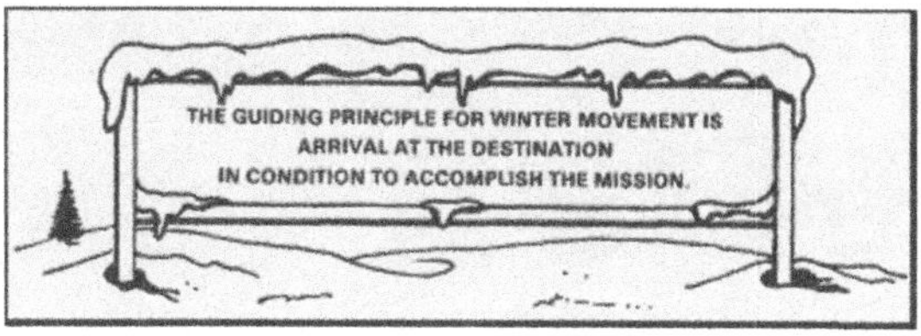

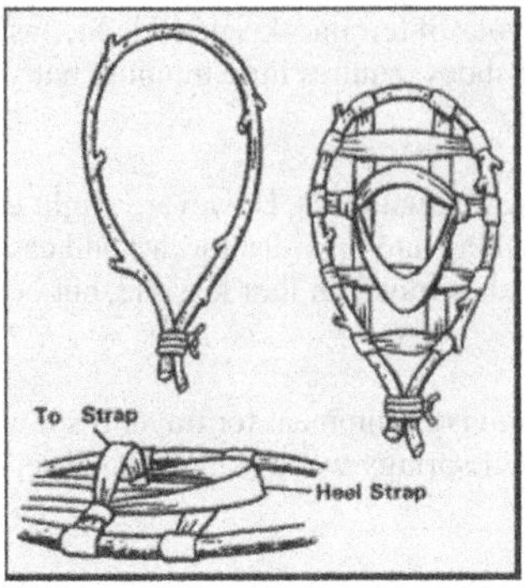


**Figure 8-2. Improvised snowshoes.**

# CHAPTER 9

# Land Navigation

***"No - I've never been lost; but for about three weeks I was kinda puzzled."***

***Brigadier General Fergusson***

## 9-1. GENERAL

In some cold-weather environments, accurate up-to-date maps may not be available and landmarks may be few or nonexistent. When the terrain is snow covered, one piece of ground looks like the other. Whatever the circumstances, you must be able to find your way — if not by map reading, by some other means of navigation.

## 9-2. CHALLENGE OF LAND NAVIGATION

Only poor road nets may exist in some areas. These will be of little use in cross-country navigation. If large-scale maps exist, they may be unreliable.

Long nights, falling or blowing snow, and fog limit visibility in winter. Tracks and landmarks may be erased by heavy snowfall. Accurately identify landmarks, through both ground and air observation, to prevent mistakes.

Conditions of daylight are different from those found in temperate regions (moderate climate). In winter there are long nights and short days.

Magnetic disturbances may be common, thereby making compass readings unreliable. Magnetic declinations may be extreme and can vary considerably at close points.

Many small lakes, ponds, and creeks will be found that are not on the map. The size and location of waterways may vary from year to year and from season to season and cannot be relied on to pinpoint a position. Large lakes and rivers can be used as important aids to navigation and movement.

## 9-4. DISTANCE AND DIRECTION

When traveling from one point to another, it is necessary for you to know the distance between them beforehand, then plot your journey accordingly.

## 9-5. MEASURING DISTANCES

### PACING

In temperate zones, pacing is the simplest method of measuring distance. In cold-weather areas, snow cover and the use of skis and snowshoes make pacing challenging, and it must be practiced repeatedly.

### USE OF FIELD WIRE

The best method for accurate ground measurement is for two men to use a piece of salvage field wire 50 meters long. The lead man marches off in the desired direction trailing the wire behind him. The second man signals by jerking the wire when the first has gone 50 meters and then moves forward to the first man's position. The first man marks the spot and continues forward for another 50 meters. This is repeated until the desired distance is reached.

### ESTIMATION

If no other method of measurement is available, the length of time spent in steady progress along a route may be used to estimate the distance traveled. The accuracy of this method depends on how well you know your marching speed over all types of terrain on foot, skis, or snowshoes.

## 9-6. DETERMINING DIRECTION

### COMPASS

The use of the compass is the most common way to determine direction. The magnetic field becomes weaker as you get nearer to the magnetic poles and can be easily affected by other objects. It is important to make sure that the compass is not affected by iron fragments in the ground, vehicles, or metal objects on you.

### THE SUN

There are two ways to determine direction by the sun:

- At noon, in the northern hemisphere, shadows fall to the north of objects; in the southern hemisphere, to the south of objects.
- Estimate direction by using a watch and the sun. In northern latitudes the hour hand is pointed toward the sun. An imaginary line halfway between the hour hand and 1200 will point south. In southern latitudes the procedure is different. Then point the 1200 toward the sun. An imaginary line halfway between the hour hand and 1200 will point north.

## NATURE

Nature will help in determining direction as in the following samples:

- Single evergreen trees, growing away from other trees, will always be more bushy on the south side.
- The bark of birch and poplar trees is whitest on the south side and darkest on the north.
- You should know from which direction the prevailing winds blow. Trees will be bent in the direction the wind is blowing or expected to blow.

## THE STARS

The North Star may be used to determine true north. This star can be easily found by its position relative to the Big Dipper. In latitudes under 70 degrees when traveling north, the North Star makes a good steering mark since its bearing is usually only 1 degree from true north and is never more than 2 1/2 degrees away. In higher latitudes above 70 degrees, the North Star is too high in the sky to indicate good direction.

A star near the horizon can be used with a bearing within 2 degrees of your compass course. However, it will only be good for up to 30 minutes when you are traveling north and only 15 minutes when traveling south. After these intervals, select another star.

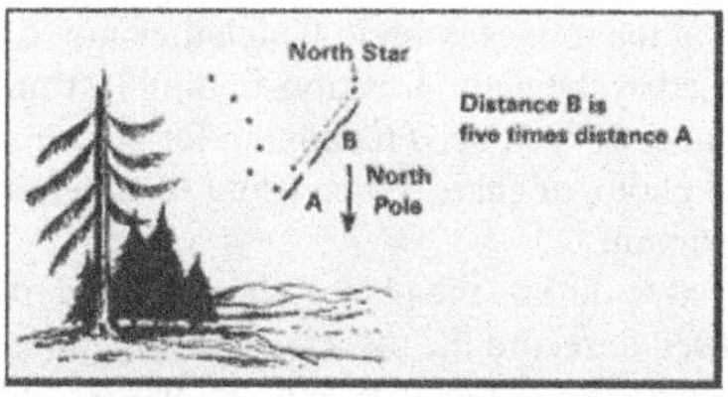


**Figure 9-1. Determining direction using North Star.**

South of the equator, the constellation Southern Cross will help you locate the general direction of south and any other direction. This group of four bright stars is shaped like a cross that is tilted to one side. The two stars forming the long axis, or stem, of the cross are called pointers. From the foot of the cross, extend the stem five times its length to an imaginary point. This point is in the general direction of south. From this point, look straight down to the horizon and select a steering point.

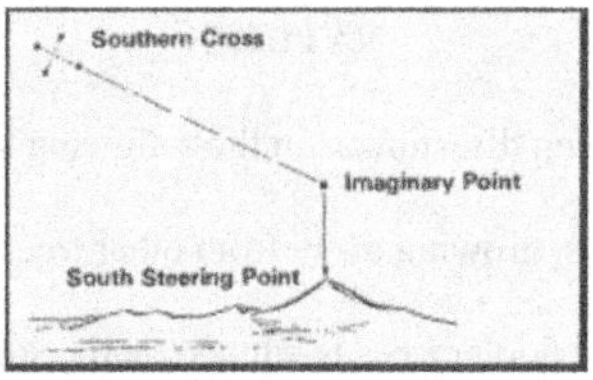


**Figure 9-2. Determining direction using Southern Cross.**

## 9-7. NAVIGATION BY DEAD RECKONING

Dead reckoning consists of plotting and recording a series of courses from a known point, each measured for distance and direction. These courses lead from the starting point to the destination. This lets the navigator determine his position at any time, either by following his plot or by comparing his actual position on the ground to his plotted course. All the details are previously estimated and recorded.

### EQUIPMENT FOR DEAD RECKONING

- Maps are used for choosing the route and for plotting the route taken as the march progresses.
- Aerial photographs are usually used in addition to maps. They must be vertical photos taken during the same season in which they are to be used.
- The compass is used as the main direction-finding instrument.
- The protractor is necessary in dead reckoning for plotting direction and distance on the map, aerial photo, or chart. When using it, remember to convert magnetic azimuth to grid azimuth.
- A route card is used to outline the plan of the proposed march. Keeping a log is mandatory. It is used to record the distance and direction.
- The log is a written record of where the soldier has started, the distance and direction traveled, and his position at any time in relation to the planned route, destination, and starting point.

### THE DRILL OF DEAD RECKONING

Know the starting point and destination and, having examined the map or aerial photo for the best route, prepare a route card. This describes each leg of the proposed route in terms of distance and direction. As the march continues, keep a careful record of each bearing taken and the distance, and record this in the log. There may be a change to the planned route because of terrain or enemy action. Therefore, the necessary changes are made using the map or aerial photo as a guide, and they are recorded in the log.

**9-8. HINTS FOR THE NAVIGATOR**

The navigator must keep the compass warm to speed up recording his position. When no steering marks exist to the front, he should march on a back azimuth. A steering mark may be a natural or man-made structure to the rear. If the steering mark is not clear, he must keep an eye on it after taking an azimuth so that it doesn't become lost. Never take magnetic azimuths in the area of metal objects.

When visibility is poor, only close-in steering marks will be available. Under these conditions try to pick up steering marks along the correct azimuth. This can be done correctly only when the route to each steering mark follows a straight line. Take compass checks often to ensure that the correct azimuth is being followed. Set the compass for a night course. A steady unshifting wind will also aid in keeping direction.

Depending on the area of the world, other signs can be used. For example, in the barren and semibarren lands of eastern Canada, snowdrifts point to the northwest and windblown trees to the southeast because of prevailing winds. In Scandinavia, scratches on rocks and cliffs caused by the glacial period (ice age) run northwest to southeast.

# CHAPTER 10

# Survival

***"Each day the number of my dogs dwindled and the sleds were broken up to cook those animals that we ate ourselves."***

***Peary, Arctic Expedition, 1906***

## 10-1. GENERAL

Your mission may take you to an arctic or subarctic region. You expect to have all personal equipment and other members of your unit with you wherever you go. But you could find yourself alone in a remote area with little or no personal gear.

## 10-2. ACTION WHEN LOST

Brief the soldier on the route before setting out on a march; he should memorize details if possible. A soldier can become separated from the main body of troops, especially if an action is fought. This applies mainly to night battle or bad weather. It can also happen if he is careless and march discipline is poor.

### WHEN LOST WITHIN A KNOWN LOCALITY

If all is quiet and there are no battle sounds or aircraft to guide you to friendly lines, go back to the last known position. If you cannot go back, consider carefully the journey already made and the distances and directions from the starting point.

Opinions should be taken from the group. Search parties should be sent out to find the next known steering mark. Trails must be marked carefully so that the search party can return to guide the main group forward or to rejoin the group, should their search be unproductive. Meanwhile, the rest of the group should seek shelter.

If the search party fails to find any known location, perform the steps below, but remember to think clearly and keep calm.

### CONDUCT WHEN LOST

When you are certain that you are lost:

- Stay together. Meet your emergency as a group; don't wander off as individuals; never allow stragglers.
- Take shelter. If you have a tent, set it up and get in it. If you do not have a tent, build one of the improvised shelters. The type of shelter is not important as long as you protect yourself from the weather.

- Keep warm. If you are lost in a forest, light a fire (depending on the enemy situation). If you are lost in an area where there are no trees, do not use your gasoline stove to heat your shelter. Use your stove only to prepare hot food and drinks. Conserve your fuel supply; you won't get any more. Get into your sleeping bag to stay warm.
- Check food supply. All the food and water in the group should be collected and rationed systematically. Supplement the food supply by shooting large animals and trapping or snaring small ones.
- Check communications. If you have a radio set, use it to contact your parent unit for help. If you are in a forest, prepare fires to attract search planes. Smoke is better by day and flame at night. Mirrors, ground signals, etc., are helpful in attracting search parties. Arrange necessary ground/air signals for search planes.
- Rest. The more you rest the less energy you expend and the less food you need.
- Adhere to survival tips. Carry safety matches inside the butt plate of your rifle or in your pockets. Carry candles, which are easy to light, provide heat and light, and help in starting a fire. If you have no matches, fire may be produced by striking the flint on your waterproof match box with the blade of a knife and collecting sparks in dry tinder. Dry moss and bark make good tinder to start your fire. Fires should be built in layers with sticks being laid out like the spokes of a wheel.
- Carry a two-cell flashlight. It can be seen from the air or on the ground as far as line-of-sight will permit.
- Ground/air signals can be made with small fires to form the identification characters (see Appendix C).
- Survival depends on clear thinking and resourcefulness. The main things to remember are to keep calm, think, try to help each other, keep together, and keep warm.

## 10-3. IMPROVISED SHELTERS

Sometimes tents or other regular shelters are not available. In summer, if the weather is mild, you may need protection only from insects. In the winter, however, you cannot stay in the open for long periods unless you are moving. In case of emergency, you must know how to protect yourself from the effects of the weather.

If natural shelters such as caves or rock ledges are available, they should be used. If they are not, a temporary improvised shelter must be built.

The type of shelter to be built depends on the equipment and materials available. By the proper use of materials available, some sort of shelter can be built during any season of the year. In open terrain a shelter can be built using ponchos, canvas, snow blocks, or other materials. Snow caves, snow trenches, snow houses, or snow holes may be built in the winter. In the woods, a lean-to is normally preferred to other types of shelter. In cold-weather areas, nature provides you with the means to prepare a shelter. Your comfort, however, greatly depends on your initiative and skill at improvising.

## PONCHO SHELTERS

The simplest type of shelter can be made by pulling the poncho over your sleeping bag. For additional comfort, various types of shelters and lean-tos can be made by attaching your poncho to trees, tree branches, or poles.

To build a one-man shelter from one poncho, spread the poncho on the ground, hood side up. The hood opening must be tightly closed by adjusting and tying the drawstrings. The poncho is raised in the middle of its short dimension to form a ridge, with the stake out at the corners and sides. Snow, sod, or branches are used to seal the side and one end of the shelter to provide additional protection from the wind and to retain heat.

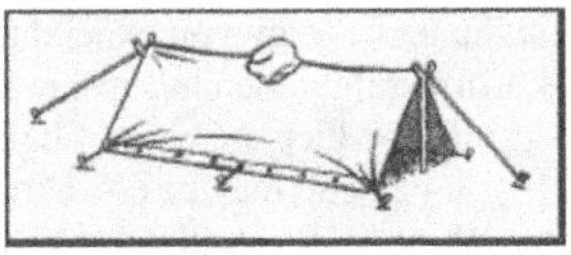

**Figure 10-1. One-man shelter.**

A two man shelter can be made from two ponchos. Spread the ponchos on the ground with the hood side up and the long sides together so that the snap fastener studs of one poncho may be snapped into those of the other poncho, The ponchos are raised where they are joined to form a ridge. The shelter is then staked out at the sides and corners. A third poncho may be snapped into the other ponchos to form a ground cloth.

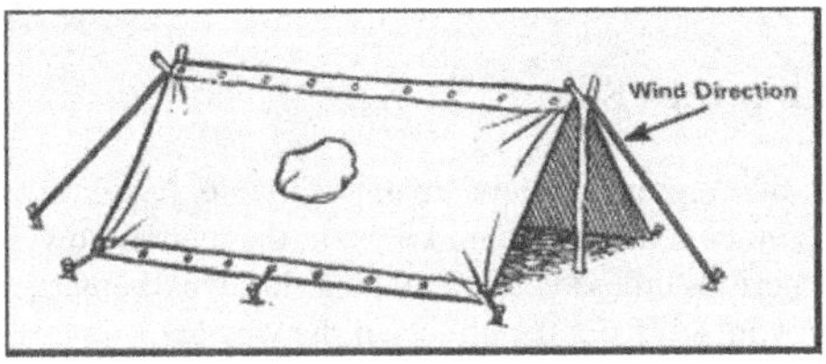


**Figure 10-2. Two-man shelter.**

## LEAN-TO SHELTER

The lean-to shelter is built in forested areas and is made of trees and tree limbs. A poncho, a piece of canvas, or a parachute, in addition to boughs, may be used for covering. Depending on the number of people to be sheltered, two types of lean-tos (single and double) are built. Boughs should be laid in the shelter like overlapping shingles (the same as on a live tree).

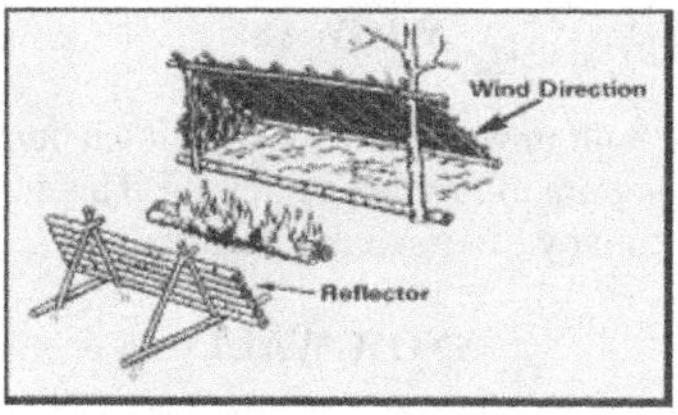


**Figure 10-3. Single lean-to.**

To save time and energy, use two trees the right distance apart and sturdy enough to support the crosspiece. If suitable trees are not available, use two forked poles or build two A-frames to hold the crosspiece. A large log is laid at the back for added height. Stringers about 3 meters long and 10 centimeters in diameter are then placed 45 centimeters apart from the crosspiece over the top of the log in the rear. Available coverings, ponchos, boughs, and so forth are then placed on top of the stringers. A double lean-to is made by building two single lean-tos facing each other with the fire in between.

Any kind of an open fire can be used to heat the lean-to. The best kind for a large lean-to is the log fire. Precautions must be taken to prevent the fire from burning too hot and burning down the shelter. A fire reflector may be built of green logs or poles to reflect the heat into the shelter and to serve as a windbreak to keep the fire burning steadily.

## TREE SHELTERS

In wooded areas, the tree-pit shelter furnishes temporary protection. Select a large tree with thick lower branches and surrounded with deep snow. Enlarge the natural pit around the tree. The walls and floor are lined with branches and boughs. Try not to disturb the snow that may be on the branches.

An emergency shelter for one man can be built by cutting down a coniferous tree at a point about 1 meter from the ground.

**Figure 10-4. Tree-pit shelter.**

After the tree has fallen, the branches on the underside are trimmed away and the cut material used for insulation on the ground.

## WIGWAM

A wigwam or tepee can be built in wooded areas by tying a number of poles near the top and spreading them at the bottom to form a large circle. This framework is then covered with ponchos, a parachute, or any other suitable material.

## SNOW WALL

In open terrain with snow and ice, a snow wall may be built for protection from strong winds. Blocks of compact snow and ice are used to build a windbreak. Be careful not to get covered up by drifting snow, which may accumulate on the lee side of the wall.

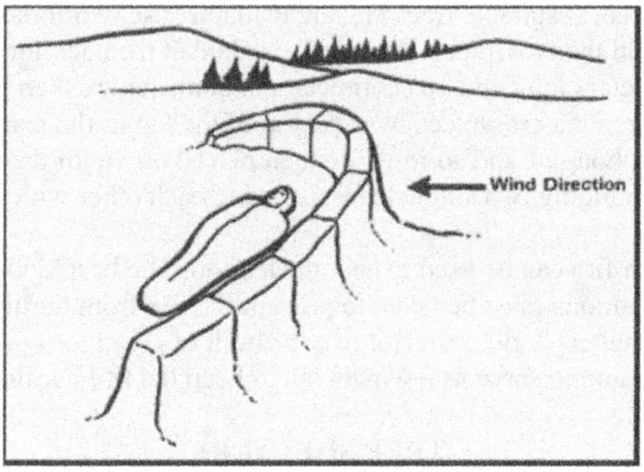


**Figure 10-5. Snow wall.**

## SNOW TRENCH

A snow trench provides shelter quickly. It is built by burrowing into a snowdrift or by digging a trench in the snow and making a roof of ponchos or ice and snow blocks supported by skis, ski poles, or snowshoes. Boughs can be used for covering the roof and for the bed. Figure 10-6 shows one method of roofing over the trench by leaning snow blocks against each other.

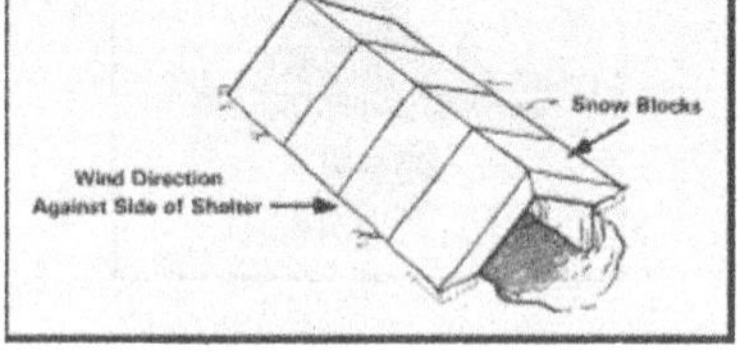


**Figure 10-6. Snow trench.**

## SNOW CAVES

A snow cave can be used as an improvised shelter in the open areas where deep and compacted snow is available. Normally, a suitable site can be located on the lee side of a steep ridge or riverbank where drifted snow collects, and it is wind packed.

The tunnel entrance must give access to the lowest level of the chamber. This is the bottom of the pit, where cooking is done and equipment is stored. The snow cave must be high enough to provide comfortable sitting space. The sleeping area must be on a higher level than the highest point of the tunnel entrance, where there is warmer air.

The roof must be arched for strength so that drops of water forming on the inside will not fall on the floor. This water will follow along the curvature of the roof and sides, glazing over the walls when frozen. The roof must be at least 30 centimeters thick. The size of the snow cave depends upon the number of men that will occupy it. A large cave is usually warmer and more practical to build and maintain than several small caves.

The shape of the snow cave can be varied to suit conditions. When the main cave is built, short side tunnels are dug to make one- or two-man sleeping rooms, storage space, latrine, or kitchen space.

In building the snow cave, locate a deep snowdrift at least 3 meters deep. Newly fallen, powdery, or loose snow should not be used. The depth of a snowdrift may be tested with a sharpened branch or young tree about 4 meters long, or with the ski or shorter ski pole. The entrance must be chosen carefully so that wind will not blow into the cave or block the entrance by drifting snow. The entrance should be built so that it is about 45 degrees from the downwind side. A small tunnel is burrowed directly into the side of the snowdrift for 1 meter. A chamber is excavated from this tunnel. Excavation is done to the right and left, so that the length of the chamber is at right angles to the tunnel entrance. Personnel doing the digging will become wet from perspiration and from the snow inside the cave. They should wear the minimum amount of clothing to ensure that they have a change of dry clothing when finished.

The cave can be heated with the one-burner gasoline stove or with candles. The fires should be put out when personnel inside the cave are sleeping, thus reducing the danger of fire, asphyxiation, or carbon monoxide poisoning. If the weather is severe and it is necessary to have a fire while sleeping, an alert fire guard must be posted in each cave. The ventilation holes must be inspected every 2 or 3 hours to ensure that they have not become clogged by snow or ice.

To ensure that the cave is warm, the entrance should be blocked with a rucksack, poncho, or snow block when not in use. All available material such as ponchos, cardboard from ration boxes, brush, and boughs should be used for ground insulation.

Walking on the roof may cause it to collapse. At least two ventilator holes are used, one in the roof and one in the door. A ski pole can be stuck through the roof ventilator to clear it from the inside. Extra care must be taken to keep air in the cave fresh when cooking or heating. The entrance should be marked by placing a pair of skis, or other equipment, upright on each side of the entry way.

Building a snow cave requires practice. Large shelters are usually more efficient for squad-size units and larger units. The most critical decision is when and where to stop and build the shelter.

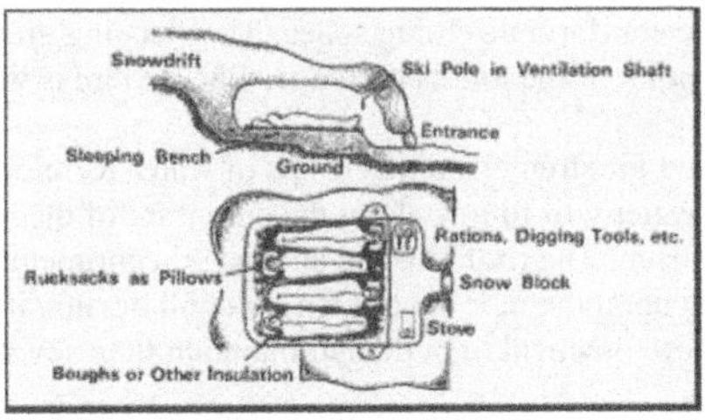


**Figure 10-7. Snow cave.**

## INDIVIDUAL SNOW HOUSE

If you are alone and have need of a shelter quickly, you can build a small individual snow house. It is box-shaped, and the only building materials needed are snow blocks. Cut snow blocks 40 by 40 by 80 centimeters. Lay them on edge to a length of 2.5 meters and a width of 1 meter. Build the walls up to a height of about 1 meter. When you have the walls built, make an A-shaped roof by leaning together pairs of blocks. Fill in the ends with snow blocks and chink all the crevices with soft snow. Cut a hole in the end of the leeward side large enough to crawl into. When you are inside, make one or two small holes near the top for ventilation. Block up the end with a snow block or with your rucksack.

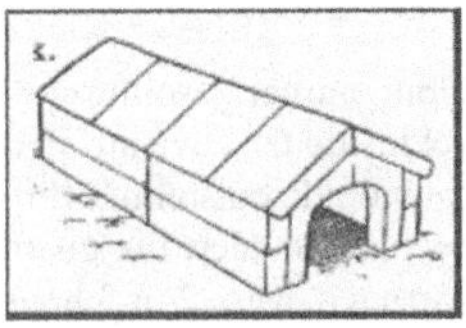

**Figure 10-8. Individual snow house.**

## SNOW HOUSE OR IGLOO

An igloo makes an excellent shelter in snow and glacier areas when other materials are not available or when the snow is so hard that it can be cut into blocks. Cracks in the igloo are then covered with loose snow. To increase the head room and to prevent water droplets from forming inside the igloo, the corners of the blocks may be cut off to form a round and smooth interior.

A shelter may be built over the entrance in the same manner as the igloo, and it should be 1 to 2 meters long and about 1 meter wide and 1 meter high. This shelter provides more protection and warmth inside the igloo and a place to store equipment. Blocks of snow or other material may be placed at both the outside and the inside entrances of the tunnel for more warmth. Make a ventilation hole near the top and at the same height as the fifth or sixth row of snow blocks. The hole can be kept open by placing a ski pole or long stick into it and clearing it out as needed. The igloo is heated with a one-burner stove or with candles.

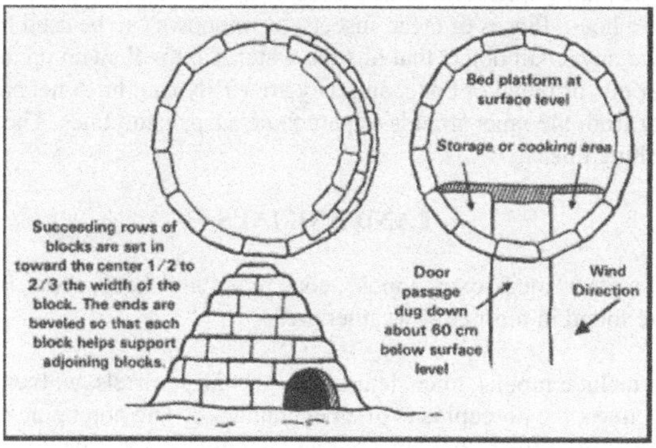


**Figure 10-9. Snow house.**

## 10-4. FOOD

The chances of finding different types of food vary in cold-weather areas depending on the time of year and the place. Shores normally are scraped clean of all animals and plants by winter ice. Food can still be found north of the timberline, even when such foods as mice, fish, and grubs are not available.

### STORAGE AND PRESERVATION

If a large animal is killed or a large number of small game are found, you should store or preserve some of the meat for future use. Freezing is the best way to preserve fresh meat or fish. It may be necessary to protect your supplies from small animals. This can be done by hanging these supplies about 2 meters from the ground.

### FISH

There are few poisonous fish in cold-weather areas, but some fish, like the sculpin, lay poisonous eggs. The black mussel may be poisonous at any season and is as deadly as strychnine. In cold regions do not eat shark meat; it is poisonous. In coastal streams and rivers, salmon moving upstream to spawn may be plentiful; however, their flesh deteriorates as they travel away from the coast, making them unfit to eat except in an emergency. A salmon that has been in fresh water long enough to turn a brilliant red is decaying and may be unfit to eat.

Coastal waters are rich in all seafood. Grayling, trout, white fish, burbot, and pike are common to the lakes, ponds, and arctic coastal plains of North America and Asia. River snails or freshwater periwinkles are plentiful in the rivers, streams, and lakes.

Fish can be speared, shot, netted, hooked, caught by hand, or stunned by a rock or club. Fish hooks can be improvised from pins, pieces of wire, or any other metal object that can be bent into a hook. Pieces of meat, insects, or minnows can be used for bait. Some fish will nibble at any small object that hits the water. Cod will swim up to investigate strips of cloth or bits of metal or bone, and they are easily caught. A net can be made out of stout twine or from the inner strands of parachute suspension lines. These strands can also serve as fishing lines.

## LAND ANIMALS

Deer, caribou, reindeer, musk-oxen, moose, elks, mountain sheep, goats, bears, and other such animals are found in most cold-weather areas.

Tundra animals include rabbits, mice, lemmings, ground squirrels, wolves, and foxes. Where there are trees, the porcupine is often encountered. The porcupine can be easily clubbed or shaken out of trees. Pick up a porcupine only after it is dead.

Hunting is better during the early morning or late evening when the animals are moving to and from feeding and bedding grounds and water. Large game such as moose and caribou should be shot just back of the front shoulder or, if the range is short and you are sure of your shot, in the neck. These large animals are fairly easy to stalk and kill, and their skins are also useful for survival. Bears should be shot just forward of the front shoulder to stop them or, if you are sure of your shot, in the neck. To successfully hunt these land animals you should know something of their characteristics:

- Caribou or reindeer are usually very curious. It is possible to attract them near enough for a shot by waving a cloth and moving slowly toward them on hands and knees.
- The technique of imitating a four-legged animal may also cause a wolf to come closer to a hunter.
- Moose may be found in heavy brush or around the shoreline of lakes.
- Mountain goats and sheep are cautious and hard to approach. They can be surprised, however, by moving quietly downwind while they are feeding. If possible, stay on higher ground than they are on.

**Figure 10-10. Hunting land animals.**

- Musk oxen leave cattle-like tracks and droppings. When alarmed, they group together in a circle with their heads facing out and remain in that position unless approached, then one or more bulls may charge.
- Bears are always dangerous. A wounded bear is most dangerous and should not be followed into cover. The polar bear is a tireless, clever hunter with good sight and an extraordinary sense of smell. Be careful that you don't become the hunted instead of the hunter.
- Rabbits often run in circles and return to the same place when they are frightened. A running rabbit can sometimes be stopped by whistling. Snares made of wire and placed on small game trails, arranged so that the animal will place his head through the loop, are an excellent means of catching rabbits.
- The quick-kill method can be used on most small animals and sometimes on larger animals after they have been shot and are on the ground. The method is simple and consists of cutting the animal's throat with a sharp knife. This system not only kills the animal but bleeds him at the same time.

**Figure 10-11. Snaring a rabbit.**

**SEA ANIMALS**

During winter and spring, sea mammals — seals, walruses, and polar bears — are found on the frozen pack ice and on floes in open water.

Seals are hard to approach but can be stalked. Keep downwind and avoid sudden moves. A white camouflage suit will help. Advance only when the animal is sleeping. Do not eat the liver of the bearded seal, because its high vitamin A content may cause you to be sick.

Walruses are found on pack ice and must be approached by boat. Because of its large size and boldness, the walrus is one of the most dangerous animals in the arctic.

Polar bears are found in almost all arctic regions but rarely appear on land. Avoid them if possible. If it is necessary to kill one for food, do not eat the liver. The liver is high in vitamin A, which may cause you to be sick. Never eat any bear meat unless it is cooked; it is always diseased.

## BIRDS

The breeding ground of many birds is in the arctic. Ducks and geese build their nests near ponds on the coastal plains during the summer and provide an important source of food. Grouse and ptarmigan live in mountainous terrain and brush-covered areas in arctic and subarctic areas. Sea birds may be found on cliffs or small islands on the coast. Sea birds, ravens, and owls are useful for food.

In winter, owls, ravens, grouse, and ptarmigans are the only birds available. Ptarmigans and grouse are easily approached and are very tame. Although they are hard to locate because of the protective coloring, they provide a good source of food because they can be killed with a stone or a club.

## PLANTFOOD

Most plants that grow in cold regions can be eaten. The water hemlock and the baneberry are the only poisonous plants. The water hemlock is one of the world's most poisonous plants. It can be recognized by where it grows, always in wet ground, and by the following characteristics: a hollow, partitioned bulb at the base of a hollow stem, spindle-shaped roots, and a foul odor, which is very noticeable in the root and bulb. The water hemlock is abundant in marshes near southern beaches and around marshy lakes in interior river valleys. It is never found on hillsides or dry ground. The baneberry is normally found on hillsides or in forested bog areas. Some mushrooms are poisonous and should not be eaten. Although the cold-region areas have many kinds of berries that are edible, most will not be available in the winter.

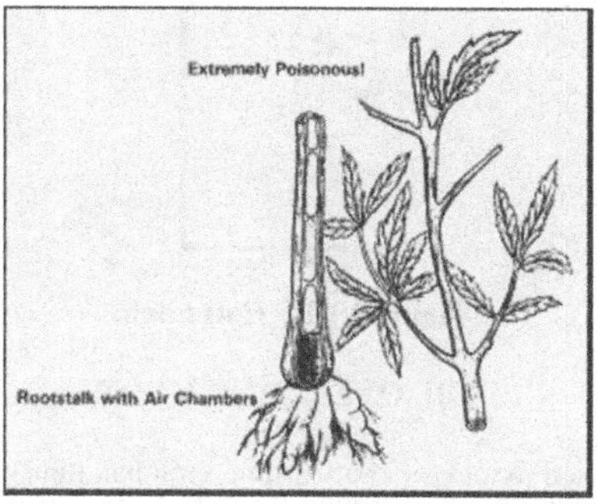


**Figure 10-12. Water hemlock.**

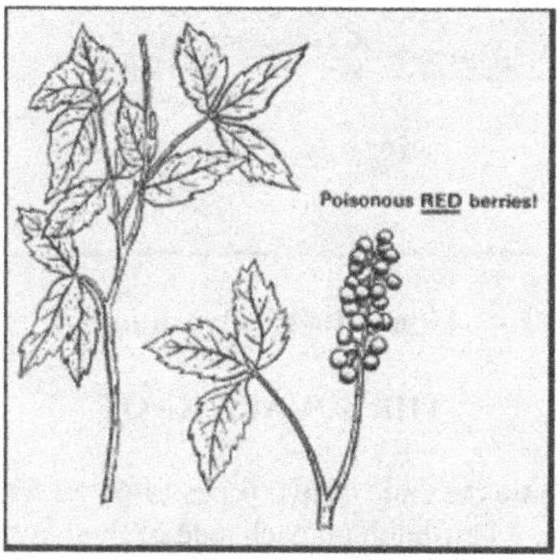


**Figure 10-13. Baneberry.**

## 10-5. USEFUL KNOTS

A rope is only a limp coil of hemp or nylon. It is useful only when you are able to attach it to itself or another object. The six basic knots described below will fit all your needs for tying rope together or onto equipment or personnel.

### THE HALF HITCH

The half hitch is used to secure the end of a rope to some other object or to secure other knots.

**Figure 10-14. Half hitch.**

## THE OVERHAND KNOT

The overhand knot is used to make a knotted rope for a handline or to temporarily whip the end of a rope to keep it from unraveling.

**Figure 10-15. Overhand.**

## THE SQUARE KNOT

The square knot is used to tie the ends of two ropes of equal width together. The knot should be secured by tying a half hitch on each side of the knot. If properly tied, the square knot will not slip or jam. Be careful to follow Figure 10-16 or you may end up with a "granny knot," which can slip and jam.

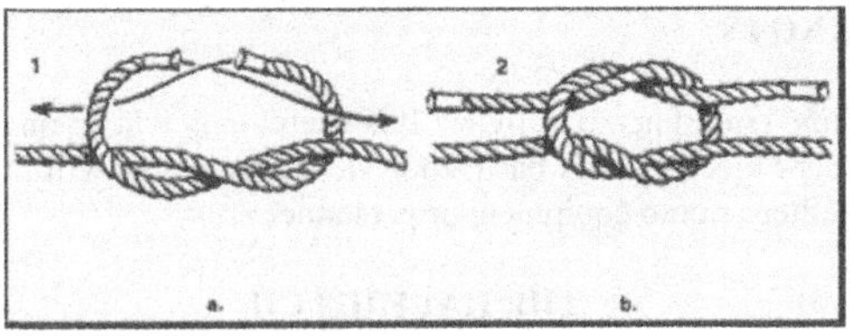


**Figure 10-16. Square.**

## THE CLOVE HITCH

The clove hitch is the most useful of the hitches. It can be used to tie the end of a rope to a post or eye, or it can be used to secure the middle of a rope without using the ends.

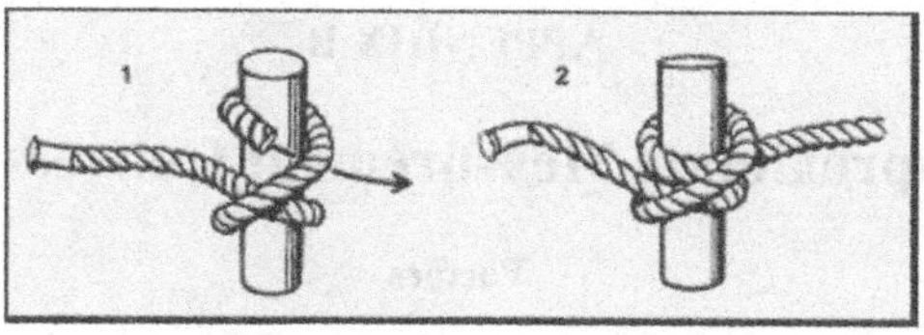

**Figure 10-17. Clove hitch.**

## THE DOUBLE SHEET BEND

The double sheet bend is used for tying two ropes together of different width.

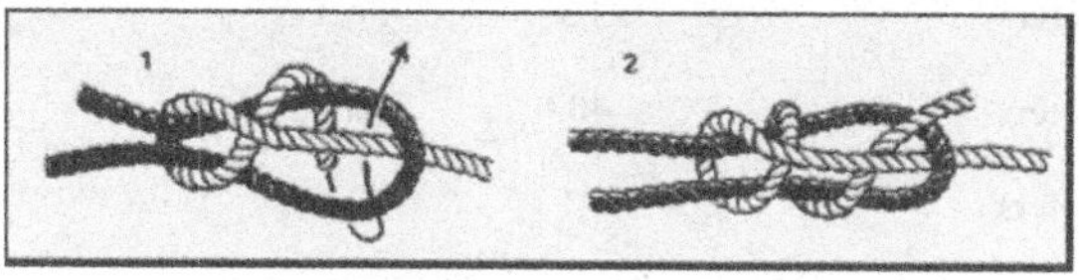

**Figure 10-18. Double sheet bend.**

## THE BOWLINE

The bowline is used to form a loop on the end of the rope that will not slip.

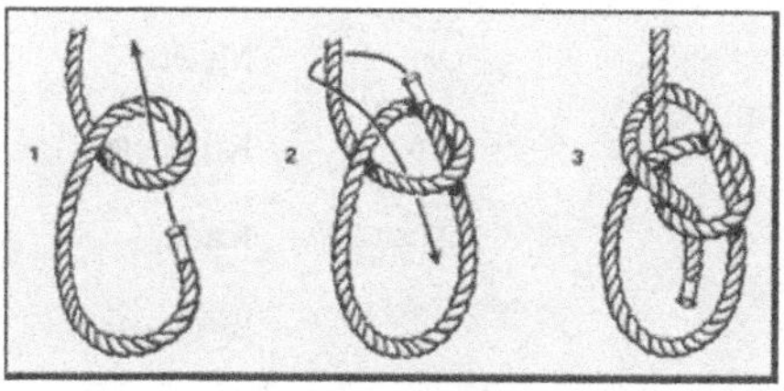

**Figure 10-19. Bowline.**

Remember the qualities of a good knot: it is easy to tie, it does not jam and become hard to untie, and it does not slip when weight is put on it.

**APPENDIX B**

# Approximate Measurement Conversion

**Factors**

| *Multiply* | *By* | *To Obtain* |
|---|---|---|
| Centimeters | .03 | Feet |
| Centimeters | .4 | Inches |
| Meters | 3.3 | Feet |
| Meters | 40.0 | Inches |
| Meters | 1.1 | Yards |
| Kilometers | .6 | Miles |
| Knots | 1.2 | MPH |
| Inches | 2.5 | Centimeters |
| Feet | 30.0 | Centimeters |
| Yards | .9 | Meters |
| Miles | 1.6 | Kilometers |
| MPH | .9 | Knots |

For ease in conversion the above factors have been rounded off.

# References

## Suggested Readings

| | |
|---|---|
| DA Pam 20-201 | Military Improvisations During the Russian Campaign |
| DA Pam 20-203 | Russian Combat Methods in World War II |
| DA Pam 20-231 | Combat in Russian Forests and Swamps |
| DA Pam 20-234 | Operations of Encircled Forces: German Experiences in Russia |
| DA Pam 20-236 | Night Combat |
| DA Pam 20-240 | Rear Area Security in Russia: The Soviet Second Front Behind the German Lines |
| DA Pam 20-269 | Small Unit Actions During the German Campaign in Russia |
| DA Pam 20-271 | The German Northern Theater of Operations, 1940-1945 |
| DA Pam 20-290 | Terrain Factors in the Russian Campaign |
| DA Pam 20-292 | Warfare in the Far North |
| FM 9-207 | Operation and Maintenance of Ordnance Materiel in Cold Weather ($0^{o}$ to -$65^{o}$F) |
| FM 31-70 | Basic Cold Weather Manual |
| FM 31-71 | Northern Operations |
| FM 90-6 | Mountain Operations |
| FM 90-11 (TBP) | Winter Operations |
| TC 90-11-1 | Military Skiing |

***"Perhaps the most valuable result of all education is the ability to make yourself do the things you have to do when it ought to be done, whether you like it or not."***

***Hosley***

# Glossary

| | |
|---|---|
| **ALICE** | all-purpose lightweight individual carrying equipment |
| **C** | Celsius |
| **CLP** | cleaner lubricant preservative |
| **ECWCS** | extended cold-weather clothing system |
| **F** | Fahrenheit |
| **FM** | field manual |
| **GP** | general purpose |
| **SFP** | synthetic fiber pile |
| **SOP** | standing operating procedure |
| **TBP** | to be published |
| **VB** | vapor barrier |